STP 1368

Concrete Pipe for the New Millennium

Iraj I. Kaspar and Jeffrey I. Enyart, editors

ASTM Stock Number: STP1368

ASTM
100 Barr Harbor Drive
West Conshohocken, PA 19428-2959

Printed in the U.S.A.

Concrete pipe for the new millennium / Iraj I. Kaspar and Jeffrey I. Enyart, editors.
    p. cm – (STP 1368)
    "ASTM stock number: STP1368."
    Papers from a conference held May 19-20, 1999, in Seattle, Washington.
    Includes bibliographical references.
    ISBN 0-8031-2621-2
    I. Pipe, Concrete--Congresses. I. Kaspar, Iraj I., 1939- II. Enyart, Jeffrey I., 1950- III.
    ASTM special technical publication ; 1368.

    TA447.C66 2000
    666'.893--dc21   00-020540

Photocopy Rights

Peer Review Policy

Each paper published in this volume was evaluated by two peer reviewers and at least one editor. The authors addressed all of the reviewers' comments to the satisfaction of both the technical editor(s) and the ASTM Committee on Publications.

To make technical information available as quickly as possible, the peer-reviewed papers in this publication were prepared "camera-ready" as submitted by the authors.

The quality of the papers in this publication reflects not only the obvious efforts of the authors and the technical editor(s), but also the work of the peer reviewers. In keeping with long-standing publication practices, ASTM maintains the anonymity of the peer reviewers. The ASTM Committee on Publications acknowledges with appreciation their dedication and contribution of time and effort on behalf of ASTM.

Printed in Scranton, PA
January 2000

Foreword

This publication, *Concrete Pipe for the New Millennium,* contains papers presented at the symposium of the same name held in Seattle, Washington, on 19–20 May 1999. The symposium was sponsored by ASTM Committee C13 on Concrete Pipe. The symposium co-chairmen were Iraj I. Kaspar, Consultant, and Jeffrey I. Enyart, ISG Resources, Incorporated.

Contents

Overview

As we reach the end of this century and the start of a new millenium we need to look at where concrete pipe has come, and also where it is going in the new millenium. While concrete pipe was in use prior to the start of the 20th century, the industry has made tremendous advancements in the last hundred years. High speed, efficient, automated plants have been developed revolutionizing manufacturing. First Dr. Anson Marston and Dr. Merlin Spangler at Iowa State University, and then more recently Dr. Frank Heger of Simpson, Gumpertz and Heger, have made tremendous advances in the technical understanding and design procedures for the internal and external performance of concrete pipe. Even with all these advances there are still many opportunities for increased understanding and improved performance for concrete pipe in the new millenium.

This Special Technical Publication has been published as a result of the May, 1999 Symposium on Concrete Pipe for the New Millenium, held in Seattle, Washington and sponsored by ASTM Committee C13 on Concrete Pipe. The objectives of this Symposium were to present historical information on the evolution of specifications and manufacturing technology for concrete pipe; to discuss innovative applications and uses; to introduce new technologies for concrete pipe products; and to both discuss and determine the use of and need for new ASTM standards for these products. This publication presents design application methods using the newly developed Standard Installation Direct Design (SIDD) methods as applied to low-head pressure pipe along with the results of installation testing and performance to verify the SIDD performance assumptions. In addition to a review of the impact of proposed load resistance factor design (LRFD) methods, developments of new technology, particularly in materials performance, is included.

Engineers will find the presentation of new design methods, and the reporting of field performance to verify these design methods, useful in advancing their understanding of current design and performance. While the information and performance opportunities using material advancements will require additional applications and performance studies, they provide an insight into the potential available with new materials. This publication just touches on some of the improved materials available now, the new millenium will bring other new innovations that will further revolutionize concrete pipe.

Iraj I. Kaspar
Consultant
Springfield, IL
Symposium Co-chairman and Editor

Jeffrey I. Enyart
ISG Resources, Inc.
Houston, TX
Symposium Co-chairman and Editor

New Technology

Leonard W. Bell,[1] William E. Shook,[2] and Troy Norris[3]

Mitigating the Corrosion of Concrete Pipe and Manholes

Reference: Bell, L. W., Shook, W. E., and Norris, T., **"Mitigating the Corrosion of Concrete Pipe and Manholes,"** *Concrete Pipe for the New Millennium, ASTM STP 1368,* I. I. Kaspar and J. I. Enyart, Eds., American Society for Testing and Materials, West Conshohocken, PA, 2000.

Abstract: This paper deals with the problems of corrosion caused by sulfuric acid generated within sewer systems. The problems are identified and potential economical solutions are presented. There are four major ways to mitigate the corrosion of concrete pipe and manholes, due to sulfuric acid produced in a sewer system:

- Utilize Az design to elevate the alkalinity of the concrete.
- Coat or line the pipe and structure.
- Reduce the microbial induced corrosion (MIC), using computer model designs.
- Use acid-resistant cements and antibacterial additives.

The last two methods will be discussed at length because they are the most cost-effective means of extending the life of concrete in a sewer system. By reducing the generation of hydrogen sulfide and at the same time reducing the microbial activity in the system, MIC is effectively reduced. Also, by incorporating acid resistant cements and antibacterial additives, concrete in sewer systems will experience less or no corrosion; thus the life of the sewer system is extended.

Keywords: Microbial induced corrosion, Thiobacillus bacteria, hydrogen sulfide, antimicrobial, concrete pipe

Environmental awareness, increased population densities, improved technology and fiscal restraint have combined to make MIC one of the major problems municipal engineers face today when designing wastewater systems. Rapidly increasing populations and population densities produce more wastewater for treatment. Our environmentally conscious society requires us to treat sewage so that it is harmless when the waste stream returns to our lakes, rivers and oceans.

This wastewater system requires a maze of piping, manholes, pump stations, and structures. Because of its strength and economy, concrete is one of the most widely used construction materials in this system. From a concrete-corrosion point of view, all these factors combine to give necessity for finding better solutions for reducing microbial induced corrosion (MIC).

In the area of wastewater design, the industry has made many advances over the last

---

[1]Director – Engineering Services, Synthetic Industries, Inc. – Fibermesh Division, 4019 Industry Drive, Chattanooga, TN 37416

[2]President, AP/M Permaform, 6250 NW Beaver, Suite 6, Johnston, IA 50131

[3]Vice President, Technical Services, Environmental Consortium, 2844 Salem Road, Conyers, GA 30013

twenty years. Pipe manufacturers now produce pipe that is much more "water tight". Very little sewage can escape out of the line and very little groundwater can infiltrate the pipe. The sewage is now more concentrated and more corrosive. Within the last decade, the ability to see inside an installed sewer pipe via remotely controlled closed circuit television has allowed engineers to actually view the results of ongoing MIC.

The current state of the infrastructure has encouraged municipalities to design their structures for maximum longevity. The Greater Houston Wastewater program represents one of the United States largest wastewater utilities [1]. Houston, according to the United States Environmental Protection Association 1992 Needs Report [2], reported that over 9,000,000 lineal feet of RCP needed to be replaced due to MIC. Currently, Houston is in the process of spending $1.9 billion to repair what is largely the result of MIC [3]. This story is repeated over and over in large and small municipalities around the world [4]. Engineers must design to combat MIC in order to increase the longevity of the sewer system and to make the system more economical and cost effective.

C.D. Parker in 1945 was one of the first to report the source of microbial induced corrosion (MIC) as the bacteria known as Thiobacillus [5]. This corrosion process is sometimes incorrectly referred to as hydrogen sulfide (H_2S) corrosion. H_2S alone is not corrosive to concrete whatsoever. It is the sulfuric acid (H_2SO_4) that is produced when the Thiobacillus bacteria metabolize the H_2S that actually corrodes the concrete. It is beyond the scope of this paper to detail the complete MIC cycle. For further information, the reader should see the ASCE Manual of Practice No. 69 [6].

When the wastewater steam is anaerobic (no oxygen is present), sulfate-reducing bacteria, existing in the slime layer in the invert of the pipe, convert the naturally occurring sulfates in the wastewater into H_2S. Numerous factors lead to greater H_2S production. It is a well-known fact that warmer temperatures result in more bacterial activity and greater H_2S production. Also, geographic regions with greater nutrients (B.O.D.) content in the water have a greater H_2S potential. The flow rate of the pipeline is a very significant factor as well. Lines with low or stagnant flows have a greater tendency to become septic and provide more anaerobic conditions for the production of H_2S. Greater flow rates help to introduce oxygen into the wastewater to prevent the system from becoming anaerobic. Higher flow rates also tend to clean away the slime layer to reduce the quantity of bacteria that can produce H_2S.

Released H_2S gas reacts with the moisture in the crown area to form dilute acids. The dilute acids reduce the pH on the surface of the concrete from its normal level of 11 or 12 to approximately pH 7 [fresh concrete pH measures approximately 12.5, but due to aging and natural carbonization, the pH level drops below 12.5 [7].

The Thiobacillus bacteria, which exists only at pH's of 7 and below, further metabolizes the excess H_2S into H_2SO_4 (sulfuric acid). Successive generations of the bacteria continue to produce the acid and lower the pH to approximately 0.9. In practical terms, the cycle maintains a sulfuric acid concentration of approximately 5% to 10%. Once the pH drops below approximately 1.25, the H_2SO_4 corrodes the concrete by reacting with the calcium hydroxide of the cement that binds the sand and aggregate together [8]. It should be noted that MIC occurs in the crown area of the pipe above the water line. If the area below the water line is corroded, it is most likely erosion caused by excessive velocities or abrasive materials in the pipe. Corrosion below the water line could be caused by other acids and chemicals in the waste stream as well.

Presentation

The first step in reducing and eliminating MIC is to design the wastewater collection and transmission systems to reduce to opportunities for H_2S production. One of the most significant design changes to occur in the last 18 years is the development of computer programs for sulfide and corrosion prediction. The most recent versions of these programs allow the user to analyze an entire system for sulfide generation and corrosion potential. When verified and calibrated, the model is a powerful tool which can be used to analyze the varying conditions anticipated throughout the life of the wastewater collection system. Using the manual method, this same analysis would require extensive time and severely limit the size of the project, which could be analyzed, and the detail of analysis, which could be performed. With a computer supported modeling technique, the model could be used as an Operations and Maintenance (O&M) tool. The impact of diversions, future flows, and changes in wastewater characteristics can all be analyzed before potentially costly decisions are made.

The most recent generation of programs published for sulfide generation and corrosion prediction are HS and Sulfide Works. Both were published in 1991. HS was developed through the American Concrete Pipe Association. Sulfide Works was developed by MicroComp Systems. Each program is provided with documentation and is based on the Pomeroy - Parkhurst Equations and the Corrosion Rate Predictive Model. The HS program is limited to pipes flowing partially full. This limitation requires manual input when modeling siphons or force mains. Sulfide Works' program handles either full-flowing pipes or partially full pipes.

When evaluating a system's sulfide potential, it may be necessary to simulate varied conditions. The programs provide various options, including constant or variable quantity or depth of flow and incremental life analysis, to account for variable flow quantities of depths during the sewer life, and will take into account the effect of input sulfide at junctions. For primary data input, sewage characteristics required are: climatic BOD, sewage temperature, design life [which may be broken into increments], acid reaction factor "k", pH of the sewage, upstream total sulfide level, insoluble sulfides, and the climatic ratio "c". The programs prompt for the number of reaches to be analyzed; then for the pipe diameter, slope and length of reach for each reach in succession, beginning at the upstream end of the sewer.

With the information provided by the software programs, and more specifically the "snapshot" information available from the ACPA Hydrogen Sulfide Prediction software, the designer can work with different "what if" scenarios to determine the best design for the wastewater system. These are important to the specific application, both at present and in the future.

Today's designer can have the modern day equivalent of a crystal ball, which allows the estimation of tomorrow's Operations, Maintenance, and Replacement (OMR) costs. H_2S Modeling Design Method software is used in estimating the future costs of wastewater systems. Pipe and all the other components of the wastewater system can be initially designed, rehabilitated or studied for future design and maintenance costs. Community expansion, real time and planned, can be accommodated by the H_2S Modeling Design Method program. Design professionals can utilize H_2S Modeling Design Method to determine future needs.

Deterioration of present systems can be determined prior to the system becoming a major problem. H_2S Modeling Design Method information is not only valuable at design time but also at rehabilitation time. Annual maintenance budgets can be accurately predicted by effectively utilizing the H_2S Modeling Design Method software on a periodic scheduled time frame. Design professionals can now have accurate input into wastewater system maintenance costs. Graphic presentations can be presented to city officials to support budget requests and to illustrate the construction and rehabilitation needs of the city's wastewater system.

As with any software of mathematical concept, H_2S Modeling Design Method is only as good as the data input. The accuracy of the data, and the skill and knowledge of the operator, are key factors in successful H_2S modeling.

Because the factors controlling sulfide generation in sewers are so complex, it would be unrealistic to expect that sulfide concentrations can be accurately predicted on an hour-by-hour basis. Even predictions of average sulfide conditions in a sewer are not considered precise, but they will be adequate for many design and operation purposes. The Pomeroy equations that have been devised have coefficients that can be modified to meet the objectives of the engineer, giving results that will approximate average performance of all sewers represented by any given set of parameters or that will give results in varying degrees of conservatism. Thus, a sulfide prediction may show a concentration that will rarely be exceeded in any sewer, or one that will be exceeded only part of the time, or one that will be an average value where septic conditions prevail.   The level of understanding of sulfide generation mechanisms and corrosion of both cement bonded and ferrous materials allows a relatively accurate assessment of anticipated conditions in sewer systems and the cost-effective design of control measures. Structures, manholes, tanks and the treatment plants can benefit from the use of H_2S Modeling Design Method evaluation. This concept allows for the selecting of methods to minimize the corrosion of all concrete and metallic elements.

By utilizing the H_2S Modeling Design Method the following major factors can be addressed to minimize the formation and presence of sulfide in sewage systems.

- Limit the use of closed conduit systems [force mains, siphons, and surcharged sewers].
- Provide for velocities in both gravity and pressure pipes that are adequate to prevent deposition and accumulation of solids, especially during periods of low flow.
- Provide velocity in gravity trunk sewers and interceptors, such that surface re-aeration is adequate to prevent sulfide build-up.
- Eliminate direct discharge of sulfide to the wastewater collection system from industrial and septic waste sources.
- Minimize the accumulation of solids in the treatment plant at any location where they will become anaerobic and septic.

In addition numerous other methods are available to control the generation of sulfide in wastewater. These methods affect the oxygen balance in sewage, oxidize generated sulfide, and react chemically with dissolved sulfide to form insoluble sulfide, or affect the sulfide generation capability of the sulfate or organic sulfur reducing organisms. The methods include: 1) Oxygen injection in force mains, inverted siphons, U-tubes, hydraulic falls, and side streams; 2) chlorination; 3) hydrogen peroxide; 4) iron and zinc salts; 5) shock dosing with sodium hydroxide; 6) potassium permanganate; 7) sodium nitrate; 8) ozone; and 9) bacterial cultures and enzymes.

The second step in reducing and eliminating MIC is to prevent the Thiobacillus from growing, thus cutting off the biogenetic formation of sulfuric acid. Traditionally, efforts to control corrosion of concrete sewers have focused on coating the concrete or using plastic liners, or chemical treatments to reduce the concentration of dissolved hydrogen sulfide gas carried by the wastewater. Most of the treatments are costly and do not provide adequate, long-term protection or control. Concrete, which is coated, is not 100% effective. Acid can penetrate coatings though pinholes and react with the concrete; thus destroying the bond of the coating to the concrete [9]. Thiobacillus are still present and able to produce sulfuric acid on the surface of the coating. By adding an antimicrobial agent to the coating some of this action can be abated. Another concern about coatings is their adhesion to the concrete. Pull off testing of coatings has shown the failure zone within the concrete because the surface concrete pulls away with the coating. By increasing the cohesion of the concrete, greater resistance to coating pull off would be achieved. It stands to reason that fibers in the concrete would improve the concrete's cohesion and thus the coatings adhesion. For improved coating performance, the use of an antimicrobial agent in the coating and better adhesion through the use of fibers in the concrete is logical.

Another approach that has shown success in mitigating corrosion is mortar made from calcium aluminate cement and same source clinker. This material has shown a reduction affect on Thiobacillus growth and greater resistance to MIC although it is still subject to corrosion, albeit at a slower rate. Very recently new cements, which are acid resistant, have arrived in the market place which exceed the requirements of ASTM Performance Specification for Blended Cement (C1157M-95).

These cements can be used in place of portland cement and do not produce calcium hydroxide, which is attacked by sulfuric acid. They are made by blending fly ash with chemicals and are presently approved for concrete production under ASTM Specification for Ready-Mixed Concrete (C94-99).

Additionally, a new concept that has come into being is to damage the Thiobacillus cell growth by the use of an antimicrobial agent. Antibacterial materials have been in use for many years in products like "Dial Soap", which contains such material. Under today's government regulation, EPA regulates all such materials to be sure they are safe and nontoxic to humans, and other high-life forms. For years these materials have been used in antiseptic soaps and lotions for skin, disinfectant for medical instruments and for food and dairy equipment. The material, CONSHIELD™[1], being reported in this paper is a stable, quaternary, ammonium-salt derivative, which was developed at Emory University in Atlanta, Georgia for medical purposes. This material is water-soluble which makes it unique as an additive for concrete. It also can be applied in liquid form to a surface where a molecular bond is established. The other interesting feature of this material is that the chemicals in it are surface-active-agents and their antimicrobial activity is directed toward the bacteria cell membrane. By disrupting the membrane, the cell can not divide and thus it will die. Cell growth is binary. If one cell takes one hour to divide, after 24 hours there will be 16,777, 216 cells produced.

With this material, testing commenced in June 1996, in the laboratories of Custom Biologicals, Inc., in Boca Raton, Florida. The investigative work was performed under

[1] AP/M Permaform, 6250 NW Beaver, Suite 6, Johnston, IA 50131

the direction of Dr. Clarence L. Baugh. Dr. Baugh is well recognized in the field of microbiology, having published works dating back to 1959 and holds patents for Interferon Production and the Production of Mareks Disease Vaccine, among others. Wafers of concrete mortar were prepared by integral mixing of the antimicrobial solution with the mortar. Other untreated wafers were coated with the same solution. These samples along with plain (control) samples were forced to a lower $pH \cong 8$ using carbon-dioxide gas to accelerate the process.

A bacterial suspension of Thiobacillus thiooxidans, Thiobacillus thioparus, and Thiobacillus denitrificans were aseptically pipetted evenly onto the surface of concrete wafers and incubated at 25°C for 24 hours. Viable counts were then obtained using a modified NETAC method (this is a method microbiologists use to determine cell counts). Four test replicates were made per set and incubated at 25°C for 26 days. All of the organisms were killed by the test material with a complete kill of 24 hours. (See Table 1) In addition to the viable counts, a pH change did not occur and no growth was detected microscopically.

Table 1 – *Thiobacillus Inoculum Test Results*

Specimen	Sample	Viable Count After 24 Hours	% Reduction
T.denitrificans	Control	1×10^7	0%
T.denitrificans	CS In	1×10^2	99.999%
T.denitrificans	CS On	1×10^2	99.999%
T.thioparus	Control	1×10^7	0%
T.thioparus	CS In	1×10^2	99.999%
T.thioparus	CS On	1×10^2	99.999%
T. thiooxidans	Control	1×10^3	0%
T. thiooxidans	CS In	0	100%
T. thiooxidans	CS On	0	100%

From past experience the authors realized that laboratory results are indicators and real results have to come from the field or a working environment. With this in mind and because the laboratory results were very encouraging, field trials were undertaken.

The next step was to prepare samples for testing in a municipal sewer system. These samples were cores taken from concrete pipe commercially produced using the antimicrobial agent as an additive to concrete mix. The test protocol called for weighing the samples in a saturated-surface dry condition, and reading the initial sample surface pH. A sewer manhole was selected, which had very obvious corrosion taking place and very high H_2S reading. The samples were suspended three (3) feet below the manhole cover and approximately seven (7) feet above the flow line, and left there for three (3) months.

The original intention was to take readings of pH and weight loss after one year. At three- (3) month's time, a visual inspection was made just to check on the samples. Because of visible deterioration of the control samples, measurements were taken immediately instead of waiting for one year. The three- (3) month readings of pH and weight loss are shown in Table 2.

Table 2 – *In-situ Sewer Manhole Field Tests*

CONCRETE SAMPLES	INITIAL		FINAL		WEIGHT LOSS (GRAM)
	Weight (GRAM)	pH	Weight (GRAM)	pH	
Core from concrete pipe	894.3	11	891.4	3	2.9
Core from concrete pipe without additive	890.8	11	860.2	1	30.6

Based on the excellent preliminary test results, the City of Atlanta now specifies this material for all new and rehab cement in their sewer system. Other cities where it has been used in manhole rehabilitation are:

- Columbus, Ohio
- Ft. Walton Beach, Florida
- Mt. Prospect, Illinois
- Corsicana, Texas

Further testing in municipal sewer manholes is being conducted by Iowa State University using the antimicrobial material to further verify the initial test results.

After the tests were conducted using the additive, another option became available with the incidence of the acid resistant blended cement, further testing of sample in the manhole was conducted. Test results are shown in Table 3.

Table 3 – *In-situ Sewer Manhole Field Tests*

CONCRETE SAMPLES*	INITIAL		FINAL		WEIGHT LOSS (GRAM)
	Weight (GRAM)	pH	Weight (GRAM)	pH	
Acid resistant cement w/CS	539.4	11	539.1	3	0.3
Acid resistant cement	543.5	9	540.0	1	3.3
Plain Portland	470.0	9	446.1	1	23.9

(*) Suspended three (3) months in high concentration of H_2S gas.

Note the high degree of resistance to MIC that the combination of blended acid-resistant cement and the antimicrobial additive produced.

Conclusion

A well operating waste water system is essential to the health and well being of any society. Because the piping system is buried in the ground, it should last forever once it is placed. Many sewer systems need upgrading because of undersize or deterioration of the piping. New technology is providing ways of doing some upgrading of existing piping, by insertion of liners, replacement by pipe bursting and shotcreting of large diameter pipe.

This paper has discussed two approaches for controlling MIC of concrete sewers.

1. New computer programs can be used to reduce the beginnings of MIC, i.e., the generation of hydrogen sulfide gas and subsequently corrosive sulfuric acid.

2. Early results with a new material, a stable quaternary ammonium salt derivative, added to concrete shows promise as a means of controlling bacterial growth on the concrete and reducing MIC.

References

[1]   Lovett, P., Municipal Coating Industry Market Study, P. D. Lovett and Company, Baltimore, MD, 1994.

[2]   EPA, Needs Survey Report to Congress, EPA 832-R93-002, United States Environmental Protection Agency, Washington, D.C., 1992.

[3]   Oradat, G., "Paper Presentation," American Concrete Pipe Association 1995 Marketing/Production Short Course, San Diego, CA, 1995.

[4]   Laughlin, J., "Top U.S. Cities Planning $24 Billion in Projects," *WaterWorld News*, July/August, Tulsa, OK, 1995.

[5]   Parker, C.D., "Mechanics of Corrosion of Concrete Sewers by Hydrogen Sulphide, Sewage, and Industrial Wastes", Vol. 23, No. 12, *Australian Journal of Experimental Biology and Medical Science*, Sydney, Australia, 1945.

[6]   "Sulfide in Wastewater Collection and Treatment Systems," *ASCE Manual of Practice No. 69*, Reston, VA, 1982.

[7]   Kosmatka, Steven H. and Panarese, Williams C., *"Design and Control of Concrete Mixtures, Portland Cement Association,"* Thirteenth Edition.

[8]   Sand, W., et al., "Tests for Biogenetic Sulfuric Acid Corrosion in a Simulated Chamber Confirms the on-site Performance of Column Aluminate-Based Concretes in Sewage Applications," ASCE Materials Engineering Conference, San Diego, CA, 1994.

[9]   Redner, J. A. Randolph, P. His, and Edward Esfandi, "Evaluation of Protective Coatings for Concrete," County Sanitation District of Los Angeles County, Whittier, CA, 1992.

A. H. Vroom,[1] Leif Aarsleff,[2] and C. H. Vroom[1]

Sulfur Concrete for Corrosion-Resistant Sewer Pipe

Reference: Vroom, A. H., Aarsleff, L., and Vroom, C. H., **"Sulfur Concrete for Corrosion-Resistant Sewer Pipe,"** *Concrete Pipe for the New Millennium, ASTM STP 1368*, I. I. Kaspar and J. I. Enyart, Eds., American Society for Testing and Materials, West Conshohocken, PA, 2000.

Abstract: Experience to date shows sulfur concrete to have many advantages over hydraulic cement concrete for sewer pipe. Its durability, high corrosion-resistance, impermeability and high mechanical strength, combined with its high volume production capability, make it an economical, long-lasting material. Other properties include high abrasion resistance and extremely high resistance to fatigue.

Sulfur concrete sewer pipe can be dry-cast, wet-cast or spun using conventional concrete pipe molds. The hot mix develops its strength simply on cooling, no curing being required.

Details of mix design and preparation are described. Examples of commercial-scale production of pipes and tanks are presented.

The need for modified ASTM Standards for such pipe and related products is discussed.

Keywords: sulfur concrete; corrosion resistant; chemical resistant; acid resistance; hydrogen sulfide resistance; durable concrete pipe; concrete tanks

Introduction

Hydraulic cement concrete (HCC) sewer pipe is widely used and performs in most areas without serious problems. However, corrosion problems have been reported in many areas of the world including the United States, Germany, Japan and Russia [1]. It is subject to internal, microbial induced corrosion of concrete (MICC), and when placed in saline soils, such as found in California, Texas and the Middle East, external corrosion can also occur.

While it is possible to coat HCC pipe with epoxy, PVC lining and other materials to improve its durability, in practice these methods are expensive and attended with problems, especially, in joints. Vitrified clay pipe and different types of PVC pipe, as well as fibre glass reinforced polyester pipe, are corrosion resistant but expensive and

---

[1]Founder/Chairman and Consultant, respectively, STARcrete™ Technologies Inc., #143, 440 - 10816 Macleod Tr. S., Calgary, AB, T2J 5N8

[2]President, LA - Trading Company, Plantagevej 14, DK - 9230 Svenstrup J, Denmark

lower in strength, making them less attractive for large diameter pipe.

An ideal sewer pipe would be made of a strong and inexpensive construction material that is highly corrosion resistant and long lasting. Sulfur concrete (SC) is such a material. When formulated properly, it can provide sewer systems and industrial conduit of extreme durability for use in most environments.

SC - a thermoplastic material - is acid-proof, salt resistant and impervious to water making it fully durable in aggressive environments where ordinary HCC would deteriorate. SC is a relatively new construction material, in which hydraulic cement and water are replaced by a modified sulfur cement and fly ash or mineral filler. It is mixed and poured as ordinary concrete with the difference, however, that at 130 - 140°C, SC is a thermoplastic material that develops its unique properties simply upon cooling - normally within a few hours. Unlike HCC, which must be protected from drying while a chemical reaction occurs, SC needs no curing.

Sewage Collection Systems

Sewer pipes, manholes, catch basins and other appurtenances made from HCC are subject to microbially induced corrosion of concrete (MICC). This occurs primarily by hydrogen sulfide (H_2S), from decaying sewage, rising into the sewer's head space and being oxidized to sulfuric acid (H_2SO_4) by sulfur-oxidizing bacteria, e.g. Thiobacilli thiooxidans [1]. The sulfuric acid migrates to the moist interior surface of the pipe or other structure where it corrodes the concrete.

This reaction is enhanced at elevated temperatures and consequently is more severe in warmer climates. In order to circumvent this corrosion problem, many sewage systems have been obliged to use more expensive vitrified clay pipe or lower strength (but lighter) plastic pipe.

Within the last few years, microbial corrosion of HCC sewer pipe and its implications on infrastructure costs have been widely studied [2, 3, 4, 5].

Dr. Wolfgang Sand, an eminent European scientist specializing in sulfur reactions, has also confirmed [6] that the Thiobacilli that cause corrosion problems in sewers made of HCC do not have any effect upon sulfur concrete under similar conditions.

Researchers of the County Sanitation Districts of Los Angeles have conducted an exhaustive evaluation of protective coatings for concrete used in wastewater collection and treatment programs [7]. They examined the products available to address the significant corrosion that occurs to concrete facilities as a result of the aerobic microbial oxidation of hydrogen sulfide to sulfuric acid and the subsequent chemical reaction of the acid with the hydraulic cement in the concrete.

The researchers examined 71 different types of protective systems, including 11 specialty concretes. One of the specialty concretes, a sulfur concrete, achieved the highest total score in the test results. Sections of sulfur concrete pipe were exposed to 10% hydrochloric acid and 5% sodium hydroxide for more than two years with excellent results. A number of coatings were tested in similar fashion but with variable results.

In their conclusions, the Los Angeles researchers commented:

"A modified sulfur polymer concrete has remained in acid service for more than two years (850 days). The acid has had no effect on the concrete. If this material can be

economically used to make reinforced concrete pipe, it would eliminate the need for coatings or liners."

A major study of MICC was undertaken recently by the University of Houston in collaboration with the City of Houston. The results of these researches are expected to be published shortly.

History of SC Development

In 1972, the senior author of this paper, in co-operation with the National Research Council of Canada and McGill University, commenced a research program aimed at overcoming the problems of durability in SC experienced by other researchers [8]. The first sulfur concrete (SC) manufacturing plant began production of precast products in Calgary, Alberta, Canada, in 1975 [9].

The key to the durability of the SC produced by this process lies in the stable, microcrystalline form of sulfur it produces [10].

With a successful 20-year performance record in a variety of aggressive environments, the SC developed in Canada* is recognized as a valuable construction material in many geographic areas. Its applications include structural elements, slabs on grade and numerous precast products.

This fast-setting, chemical-resistant concrete has been used widely in the mineral processing, fertilizer, and chemical production industries due to its ability to withstand attack by acids and other aggressive chemicals [11, 12].

Since 1984, the American Concrete Institute, through its Committee 548D (Sulfur Concrete), has been developing guidelines for the use of SC [13]. The same committee is currently finalizing a "State-of-the-Art Report on Precast Sulfur Concrete" to be published shortly.

ASTM's Committee C03 on Chemical-Resistant Materials has also been actively developing standards for SC and has published the following standard specifications and test methods on the subject:

C 1159 - Standard Specification for Sulfur Polymer Cement for Use in Chemical-Resistant, Rigid Sulfur Concrete

C 1312 - Standard Practice for Making and Conditioning Chemical-Resistant Sulfur Polymer Cement Concrete in the Laboratory

C 1370 - Standard Test Method for Determining the Acceptability of Aggregates for Use in Sulfur Polymer Cement Concrete

---

* In 1973, Sulfurcrete Products Inc., now STARcrete™ Technologies Inc., of Calgary, Alberta, was established to pursue the commercial development of a sulfur concrete named "Sulfurcrete", now renamed "STARcrete™". This process uses a proprietary composition (STX™ modifier) to modify the sulfur to produce a cement comprised principally of orthorhombic sulfur in a stable, *microcrystalline* form.

Properties of SC

One of the important characteristics of SC is its extreme corrosion resistance. It is unaffected by salt, strong acids and mild alkalis (up to 5% sodium hydroxide). When made with acid-resistant aggregates such as granite and other siliceous materials, it is unaffected by continuous exposure to hydrochloric acid and to sulfuric acid up to 98% concentration.

Another characteristic of importance for many applications is its impermeability. SC repels water penetration because both sulfur and the STX additive are hydrophobic. The particles of aggregates are coated with these substances and the normal voids between particles are essentially filled with them. Water absorption after 24 hours of immersion in water at 20°C (68°F) is generally less than 0.2% by weight when measured on thin slices cut to expose the aggregates on both sides.

SC does not support combustion. Sulfur present in the surface will slowly burn when exposed to direct flame but it self-extinguishes when the flame is removed. Flame spread tunnel tests conducted in accordance with Underwriters Laboratories of Canada test S-102.2-1977 showed SC to have zero flame spread, zero fuel contribution and very low smoke density. The fire resistance is primarily due to the presence of the STX additive, which forms a protective char when exposed to direct flame. This char, together with the low thermal conductivity of sulfur, results in a relatively slow penetration of heat.

Material Comparisons - SC vs. HCC

Table 1 compares the physical strength properties of a typical SC and HCC made with the same aggregates. It will be noted that SC possesses high mechanical strength, somewhat higher modulus of elasticity (although this can be varied with additives) and higher abrasion resistance.

Water Permeability

HCC has a measurable degree of water absorptivity due, in part, to its open, continuous cell structure.

SC, however, is impermeable to water because of the hydrophobic nature of sulfur and the fact that internal cells are closed and discontinuous. Using a Hassler holder, an attempt was made to force water through a 50-mm (2-in.) thick sample at 1.03 MPa (150 psig) pressure. After 72 hours, there was zero penetration and the water absorption was only 0.46% [14].

Fatigue Resistance

SC also exhibits much improved fatigue resistance. Lee and Klaiber have shown that SC exhibits fatigue properties drastically different from those of conventional hydraulic cement concrete [15]. Beams were subjected to one and two million repetitive loadings at 90-95% of their modulus of rupture without failure. There appeared to be an endurance limit at 85-90% modulus of rupture, below which SC should exhibit no fatigue.

It is well-recognized that HCC is subject to fatigue on repetitive loading in excess of

50-55% of its modulus of rupture.

Table 1 - *Comparison of Physical Test Results*

Test*	SC	HCC
	MPa (psi)	MPa (psi)
Compressive strength	62.0 (9 000)	34.5 (5 000)
Tensile strength	7.45 (1 080)	2.6 (380)
Modulus of rupture	12.7 (1 850)	3.65 (530)
Modulus of elasticity	$3 - 4 \times 10^4$ ($4 - 6 \times 10^6$)	$2.8 - 3.7 \times 10^4$ ($3 - 4 \times 10^6$)
Linear coefficient of expansion /°C (/°F)	8.3×10^{-6} (4.6×10^{-6})	8.3×10^{-6} (4.6×10^{-6})
Density	(150 lb/ft^3)	(150 lb/ft^3)
Amount of binder	297 kg/m^3 (500 lb/yd^3)	371 kg/m^3 (625 lb/yd^3)

* ASTM standard test methods were used where applicable.
  Test results of SC are compared with those of typical HCC, both using
  19-mm (3/4-in.) washed gravel aggregate with approximately 60%
  fractured face. The relationship between compressive strength and
  modulus of elasticity can be varied with special additives, if desired.

Corrosion Resistance

SC offers extreme resistance to corrosion from acids and other chemicals whereas
HCC is notoriously weak in this respect. Fig. 1 illustrates this for sulfuric acid.
A vat, 11m long by 4 m wide and 2.5 m deep, was built using SC to hold hot sulfuric
acid. This vat was successfully used for 10 years before it was removed from service due
to a process change.

Abrasion Resistance

Using the ASTM C944 test method, the loss of weight during abrasion testing of SC
specimens, was less than half that obtained on similar HCC specimens [16].

Environmental Advantages of SC

-   no water is required in the production of SC
-   energy requirements for SC are substantially less

CO$_2$ production (air pollution) is significantly less with SC

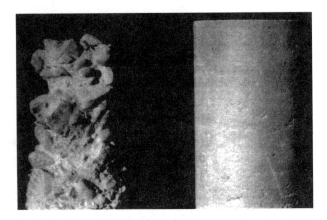

Sulfate-resistant HCC after   Sulfur concrete after
3 weeks in 20% H$_2$SO$_4$ 3 years in the same acid

Fig. 1 - *Immersion of core samples in acid*

Preparation of the SC Hot Mix

SC is normally produced in a modified asphalt batch plant. The process involves first drying and heating the aggregates to a temperature of 130 - 140°C. The hot aggregates, sulfur (either liquid or solid), STX and mineral filler are then combined in the heated mixer in proportions such as those shown below.

Pipe Reinforcement

For reinforced pipe, conventional steel rebar or wire mesh is generally used. Due to the very low water absorption of SC, corrosion of embedded steel does not occur unless the product has become cracked to permit water to penetrate to the steel.

Fibrous reinforcement, such as chopped glass fibre, also may be incorporated in the mix to increase tensile strength, flexural strength and impact resistance. Common E-glass can be used instead of the alkali-resistant glass fibre required for HCC.

Conventional prestressing and post-tensioning techniques can also be used.

SC Pipe Production Methods

Early Pipe Production with SC

Initial precasting of SC sewer pipe was conducted in the late 1970s using conventional wet casting molds meeting ASTM physical standards for 10-in. and 14-in.

i.d. pipe. Strength testing of these pipes conducted by the City of Calgary, Alberta, showed that the strengths exceeded ASTM standards by approximately 100%.

In the next phase of development (1981 - 82), 30-in. i.d. x 10 ft. sewer pipe was produced from "dry mix" SC in a commercial-scale pilot plant in California by the centrifugal, or spinning method. Using HCC, the plant was capable of making one such length of pipe every 24 hours, allowing for steam curing before opening the mold to remove the product. Using the SC hot mix and a water spray on the outside of the spinning mold, the molds could be recycled once per hour increasing production rates by 24 times. Again, physical strengths of the SC pipe greatly exceeded those of HCC pipe from the same molds.

A method for producing dry-cast SC pipe on conventional core-vibration type equipment was developed in Denmark about 1990 and patented [17] by one of the co-authors. Commercial-scale production of SC pipe of 70-cm (28-in.) diameter and circular tank sections was then initiated. Tanks in diameters up to 3500 mm (138 in.) were produced. Several tank systems with a height of 6000 mm (20 ft.) for ferrosulfate solutions at purifying plants were built. These tanks are still performing well after approximately 8 years.

Economics of Pipe Production

While the cost of the hot mix may be slightly higher for SC pipe than HCC pipe in many locations, the production cost will invariably be lower because no curing is required and pipe can be shipped the same day. For spun pipe, spinning plant capacity can be increased up to 24 times with SC as compared to HCC.

In estimating the costs of different end products, it is important that comparisons be made on the basis of life cycle cost analyses [18]. Singh [19] has reported that the U.S. Congress has passed a law requiring that all proposals for National Highway System projects, above a certain size, include a life-cycle cost analysis. As public sector funding for utilities increasingly comes under scrutiny and sanitary sewers are usually the most expensive utilities to replace, it makes sense to use the longest lasting, competitive materials that are available. ASTM C1131-95 "Standard Practice for Least Cost (Life-Cycle) Analysis of Concrete Culvert, Storm Sewer and Sanitary Sewer Systems is also pertinent in this regard.

Mix Design for Sulfur Concrete for Pipe Production

Raw Materials for SC

Sulfur
- elemental (byproduct from oil and gas refining or from natural sources)
- purity may be 70% plus, providing the clay content is less than 3%

STX
- a proprietary sulfur modifier

Aggregates
- coarse and fine, with clay content lower than 1%

Mineral filler
- fly ash, silica flour or crusher dust (used strictly to fill voids between particles of sand and control the viscosity of the liquid sulfur)

The composition of sulfur concrete for pipe production can be adapted from the composition of HCC for zero slump mixtures. The volume of hydraulic cement, filler and water can be replaced by a substantially equivalent volume of sulfur, STX and filler, while the amount of other components remains essentially unchanged.

An approximate raw material mixture for dry-casting of pipe by core vibration or spinning is shown below. This will vary to some extent with the percentage of fractured face in the aggregates and their densities.

	Weight %	Kg/m^3
Sulfur cement (sulfur & STX)	9.8	242
Coarse & fine aggregates + mineral filler	90.1	2 228
Total	100	2 470

Summary and Conclusions

- SC is not affected by hydrogen sulfide or sulfuric acid generated by microbiological activity in many sewer systems.

- SC is an impermeable, high strength, acid and salt-resistant concrete well suited for use in severely aggressive environments.

- The technology for the production of SC pipe using standard equipment is available. For concrete pipe manufacturers, it provides an opportunity to compete against other corrosion-resistant products.

- SC pipe is economically advantageous, particularly when considered using life cycle cost analysis. It may reduce maintenance, rehabilitation costs and lengthen the useful life of sewer pipes.

- There is a need for modification of ASTM concrete sewer pipe specifications to include this new material. Contact with appropriate ASTM committees must be made for this purpose.

References

[1] Padival, N. A., Weiss, J. S., and Arnold, R. G.. "Control of Thiobacillus by Means of Microbial Competition: Implications for Corrosion of Concrete Sewers", *Water Environment Research*, 1995,Vol. 67, pp. 201-205.

[2] Mori, T., Nonaka, T., Tazaki, K., Koga, M., Hikosaka, Y., and Noda, S., "Interactions of nutrients, moisture, and pH on microbial corrosion of concrete sewer pipes", *Water Research*, 1992, Vol. 26, pp. 29-37.

[3] Flemming, H-C., "Eating Away at the infrastructure - the heavy cost of microbial corrosion", *Water Quality International*, 1995, pp. 16-19.

[4] Schindewolf, J., Barnes,W.L., Kahr, C.A; Ivor-Smith,D., "A Texas-Sized SSO Solution", Civil Engineering, December 1995, pp. 55-57.

[5] Davis, J., Nica, D., and Roberts, D. J., "Microbial Populations in Sewer Pipe Corrosion Product", CSCE-ASCE Environmental Engineering Conference, Edmonton, Alberta, Canada, July, 1997.

[6] Sand, W., Univ. Hamburg, private communication, 23 June 1994, (copy available from senior author on request).

[7] Redner, J.A., Hsi, R.P., Esfandi, E.J., "Evaluation of Protective Coatings for Concrete", County Sanitation Districts of Los Angeles County, Whittier, California, Feb. 1994, 42 pp

[8] Beaudoin, J.J., and Sereda, P.J., "Freeze-Thaw Durability of Sulphur Concrete", *Building Research Note*, June 1974, No. 92, Div. of Bldg. Res., National Research Council, Ottawa, Canada.

[9] "Sulphur Concretes Go Commercial", *Sulphur Inst. J.,* Summer 1976, Vol. 12 No.2, J. Platou, Ed., The Sulphur Institute, 1140 Connecticut Ave., N.W., Washington, D.C.

[10] Gannon, C.R., Wombles, R.H., Hettinger, W.P., and Watkins, W.D., "New Concepts and Discoveries Related to the Strength Characteristics of Plasticized Sulfur", *ASTM STP 807*, American Society for Testing Materials, West Conshohocken, PA, 1983

[11] Vroom, A. H., "Sulfur Concrete Goes Global", *Concrete International,* Jan. 1998, Vol. 20, No. 1.

[12] Vroom, A.H., "Sulfur Concrete for Precast Products", *Concrete International,* Feb. 1988, Vol. 20, No. 2.

[13] "Guide for Mixing and Placing Sulfur Concrete in Construction", Reported 1993 by ACI Committee 548D, *ACI 548.2R-93,* American Concrete Institute, P.O. Box 9094, Farmington Hills, MI 48333

[14] Chemical & Geological Laboratories Ltd., Calgary, AB, Canada, Lab Report No. C80-8358, 25 Mar. 1980.

[15] Lee, D.Y. and Klaiber, F.W., "Fatigue Behavior of Sulfur Concrete", New Horizons in Construction Materials, Envo Publ. Co. Inc., Lehigh Valley, PA, 1976, p.363.

[16] Dow Chemical USA, letter report to senior author, Oct. 23, 1984.

[17] Larsen (Aarsleff), L., U.S. Patent 4,981,740, 1 Jan. 1991.

[18] Ishii, K., "Life-Cycle Engineering Design", *Journal of Mechanical Design*, 1995, Vol. 117, No.3, 42-47.

[19] Singh, B., "How Significant is LCCA?", *Concrete International*, Oct. 1996, Vol.18, No.10, pp. 59-62

Design and Installation

Frank J. Heger[1]

SIDD Installation and Direct Design for Reinforced Concrete Low-Head Pressure Pipe

---

Reference: Heger, Frank J., "SIDD Installation and Direct Design for Reinforced Concrete Low-Head Pressure Pipe," *Concrete Pipe for the New Millennium, ASTM STP 1368*, I. I. Kaspar and J. I. Enyart, Eds., American Society for Testing and Materials, West Conshohocken, PA, 2000.

Abstract: This paper describes the principal technical provisions in a proposed ASCE Standard Practice for the Direct Design of Buried Reinforced Concrete Low-Head Pressure Pipe using the same new standard installations (SIDD) as recommended in ASCE 15-93, Standard Practice for Direct Design of Buried Precast Concrete Pipe Using Standard Installations (for reinforced concrete gravity flow pipe). The proposed standard practice defines installation requirements and structural analyses for moment, thrust, and shear produced by external loads that are the same as given in ASCE-15. It incorporates design procedures for combined circumferential tension produced by internal pressure and flexure produced by external loads that are essentially in compliance with both ASTM C361 and AWWA C302 and as recommended in AWWA Manual M9. These provisions and the cited referenced standards use load factors that are larger than those used in various standards for direct design of gravity flow pipe and they further limit the maximum combined stress in the reinforcement by inclusion of a limiting maximum design yield strength of 40,000 psi, regardless of the actual yield strength of the reinforcement. There are further limits on the reinforcement stress produced by internal pressure alone as well as on the tensile stress in the uncracked concrete pipe wall that are essentially the same as given in the above standards for low-head pressure pipe.

In addition to providing the requirements and structural effects of SIDD installations, the proposed standard fills a gap in the existing standards for design of low-head concrete pipe by incorporating rational methods for calculating radial tension and shear strengths. Radial tension strength is required to resist the bending moments at the crown and invert produced by external loads. The presence of internal pressure does not increase the required radial tension strength, and the same methods used for gravity flow pipe may be used for determining the radial tension strength of low-head pressure pipe. Shear strength is another matter. The circumferential tension produced by internal pressure produces a significant effect on shear strength. Research by Dr. M. P. Collins and coworkers in Canada is the basis for new more rational procedures for shear strength of flexural members in the AASHTO LRFD Highway Bridge Specification. These are adopted and simplified for use over the range of wall thickness and strain variations applicable to low-head pressure pipe so that for the first time a rational procedure is available in the proposed standard to calculate the shear strength of low-head concrete pipe. Application of the new procedure to pipe designs for the higher fill heights combined with the highest pressure limits in the design tables of ASTM C361 indicates an adequate factor of safety for those designs, including adequate shear strength.

---

[1]Senior Principal, Simpson Gumpertz & Heger Inc., 297 Broadway, Arlington, MA 02474.

Keywords: concrete pipe shear strength, concrete pipe radial tension strength, concrete pipe standard installation, reinforced concrete low-head pressure pipe

Nomenclature

a   depth of compressive rectangular stress block produced by combined factored bending and thrust, in.

A_{sf} area of inner cage reinforcement required for flexure and thrust without internal pressure using radial tension, or flexural compression load factors, in.^{2}/ft (cm^2/m)

A_{sfmax} maximum or limiting value of A_{sf} that can be developed by radial tension strength, or flexural compression strength, in.^{2}/ft (cm^2/m)

A_{si} area of total inner cage reinforcement provided in length b, in.^{2}/ft (cm^2/m)

A_{s1} area of reinforcing near bending tension surface of wall section required for combined external load and internal pressure, in.^{2}/ft (cm^2/m)

A_{s2} minimum area of reinforcing near bending compression surface of wall section required for combined external load and internal pressure, in.^{2}/ft (cm^2/m)

b   width of section that resists stress, in. (mm)
                 taken as 12 in. [English units]
                 taken as 1000 mm [SI units],

d   distance from compression face to centroid of tension reinforcement, in. (mm)

d_s distance between centroids of inner and outer lines of reinforcing, in. (mm)

d_v 0.9d, or d_s in. (mm)

D_i inside diameter of pipe, in. (mm)

f_c' design compressive strength of concrete, lb/in.^{2} (MPa)

f_{ct} service load limit of concrete tensile stress caused by internal pressure only psi (MP$_a$)

h   pipe wall thickness, in. (mm)

f_y design yield strength of reinforcement, lb/in.^{2} (MPa)

F_c factor for effect of curvature on diagonal tension (shear) strength in curved components

F_d factor for crack depth effect resulting in increase in diagonal tension (shear) strength with decreasing d

F_{ex} factor for magnitude of strain that affects shear strength based on aggregate interlock in diagonally cracked sections

F_{rt} factor for pipe size effect on radial tension strength

f_s service load limit of reinforcing steel tensile stress caused by internal pressure only, on cracked section, psi (MPa)

g'   coefficient for depth of compressive stress block at ultimate flexural compressive strength

H   design height of earth above top of pipe, ft (m)

HAF   Horizontal Arching Factor

M_{uf1} factored moment, based on the maximum moment load factor for the flexure (condition 1) and is taken as positive for reinforcement design on the side where bending causes tension, in.-lb ft. (N-mm/m)

M_{uf2} factored moment, based on the minimum moment load factor for flexure (condition 2) for minimum reinforcement design on the side where bending causes compression, in.-lb/ft (N-mm/m)

M_{uv} factored moment caused by external loads and weight of fluid at section of critical shear in 12.9.1 or maximum shear in 12.9 2, based on load factor for shear, in.-lb/ft (N-mm/m)

M_{nuv} factored moment as modified for effects of compressive or tensile thrust, based on load factor for shear, in.-lb/ft (N-mm/m)

M_{ur} factored moment caused by external loads and weight of fluid at section of maximum moment, based on load factor for radial tension or for concrete compression, in.-lb/ft (N-mm/m)

N_p tensile thrust produced by operating internal pressure, (negative), lb/ft (N/m)

N_{pt} tensile thrust produced by operating plus transient (surge) internal pressure, (negative), lb/ft (N/m)

N_u factored thrust, acting on length b (+ when compressive, - when tensile), determined using the tensile thrust load factor for tension produced by fluid pressure and the compressive thrust load factor for compression produced by external load, lb/ft (N/m)

N_{upv} factored tensile thrust produced by internal operating plus-transient pressure above crown of pipe and weight of water (-as tension) based on load factor for shear, lb/ft (N/m)

N_{ure} factored compressive thrust at section of maximum moment produced by external load (+ as compression), based on load factor for radial tension or compressive thrust, lb/ft (N/m)

N_{uve} factored compressive thrust at section of maximum shear produced by external load (+ as compression), based on load factor for compressive thrust, lb/ft (N/m)

p_d design operating internal pressure, psi (MPa)

p_t design additional transient (surge) internal pressure, psi (MPa)

r radius to centerline of pipe wall, in. (mm)

r_s radius of the inside reinforcement, in. (mm)

V_c shear strength provided by concrete without stirrups in length b, lb/ft (N/m)

VAF Vertical Arching Factor

ϕ_f strength reduction factor for flexure

ϕ_v strength reduction factor for shear

ϕ_r strength reduction factor for radial tension

θ_v approximate inclination of diagonal tension crack, degrees

ϵ_{xu} strain in reinforcement produced by factored moments, thrusts and shears at section of maximum shear, in./in.

In 1993 ASCE promulgated ASCE 15-93, Standard Practice for the Direct Design of Buried Precast Concrete Pipe Using Standard Installations (SIDD). This is a new standard for design of reinforced concrete gravity flow pipe that incorporates two advances in the technology of buried concrete pipe: new standard installations based on more rational and quantitative geotechnical requirements and direct structural design procedures for concrete pipe in the installed design condition. These new installations are called SIDD (For Standard Installation Direct Design). For the past several years the ASCE has been developing a similar standard for buried concrete low-head pressure pipe. This paper will describe the principal features of the proposed new design practice standard for reinforced concrete low-head pressure pipe.

Concrete low-head pressure pipe has been designed based on either ASTM C361, Specification for Precast Reinforced Concrete Low-Head Pressure Pipe, or AWWA C302, Standard for Reinforced Concrete Pressure Pipe, Noncylinder Type for Water and Other Liquids, together with provisions in AWWA M9, Manual for Concrete Pressure Pipe. These standards do not contain definitive recommendations for design of standard installations for concrete pipe. Their flexural design criteria limit tensile stress (and thus, strain) produced by internal pressure and flexure from external load. However, their design procedures are incomplete because they do not include criteria for determining the pipe's capability to resist shear (diagonal tension) or radial tension produced by flexure. The latter types of behavior sometimes govern the strength of pipe under high earth covers or severe installation conditions. Thus, the principal rationale for developing the proposed ASCE Standard Practice for Direct Design of Buried Precast Concrete Low-Head Pressure Pipe is to facilitate the design of low-head pressure pipe using the SIDD standard installations and to provide criteria for shear and radial tension strength of this class of concrete pipe along with the already accepted flexural and tensile criteria.

Standard Installations

Four new standard installation types for precast concrete pipe are defined in ASCE 15-93. See also [1] for a description of these installations. Type 1 offers the highest quality of soil materials and soil compaction in the embedment zone below the pipe springline. It also requires the greatest level of field control and inspection to assure that the required support condition is actually achieved. Type 2 has been considered to be equivalent to some of the previously specified B beddings [2]. Type 3 installations permit the use of many non-plastic native soils and moderate levels of compaction. Type 4 installations require little field control and are equivalent to the previously specified D beddings [2]. The Type 4 installations utilize the inherent strength of the pipe with little help from the soil to resist external load as well as internal pressure. This type installation is appropriate for relatively shallow earth covers and for conditions where good field control cannot be obtained. External load effects that produce governing shear and radial tension designs are often found with Type 4 installations. The availability of the full range of the four standard installations enable the pipe and installation designer to select from a range of installation qualities and costs using native or imported soils that permit optimization of the combined cost of the installation and pipe.

Overall Design Procedure

The overall structural design process for reinforced concrete pressure pipe involves four basic steps:

1.   Design the soil-pipe installation. Determine the installation characteristics and the maximum static and transient internal pressures.

2.   Perform soil-pipe interaction analysis. Determine the vertical and lateral earth loads and pressure distribution on the exterior circumference of the pipe. These are given in ASCE 15 for the standard SIDD installations.

3.   Perform a structural analysis of the pipe, subject to the design load and external pressure distribution for the selected installation type, to determine the moments, thrusts, and shears around the pipe circumference produced by the design loads.

4.   Design the pipe. Select pipe wall thickness, determine area of flexural/tensile reinforcement, check for adequate shear and radial tension strength without shear and radial tension reinforcement and, if required for shear or radial tension strength, determine area and spacing of stirrups (ties). The criteria and procedures given in this paper may be used for checking shear and radial tension strength, and the procedures for designing stirrup rein-

forcement given in ASCE 15-93 may be used if shear or radial tension resistance is not adequate without radial reinforcement.

External Loads

The principal external load is the earth load. The SIDD standard installations are assumed to be subject to the earth loads and their arching factors as given in ASCE 15-93 for embankment installations. The total earth load on the pipe per unit length is determined from the weight of the column of earth directly over the outside diameter times the vertical arching factor (VAF) for the specified installation type. The VAF and the earth pressure distribution are given in Fig. 1 for each standard installation type. The relative magnitude of lateral earth load is also given in this figure as the horizontal arching factor (HAF). This figure is taken from Fig. 3 in ASCE 15-93. The standard embankment installation earth loads and pressure distributions may be used for trench installations without the necessity to control the maximum width of the trench in the field.

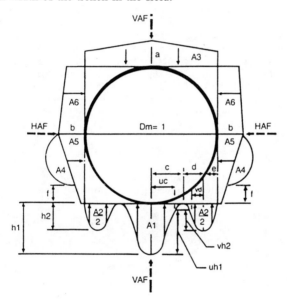

Installation Type	VAF	HAF	A1	A2	A3	A4	A5	A6	a	b	c	e	f	u	v
1	1.35	0.45	0.62	0.73	1.35	0.19	0.08	0.18	1.40	0.40	0.18	0.08	0.05	0.80	0.80
2	1.40	0.40	0.85	0.55	1.40	0.15	0.08	0.17	1.45	0.40	0.19	0.10	0.05	0.82	0.70
3	1.40	0.37	1.05	0.35	1.40	0.10	0.10	0.17	1.45	0.36	0.20	0.12	0.05	0.85	0.60
4	1.45	0.30	1.45	0.00	1.45	0.00	0.11	0.19	1.45	0.30	0.25	0.00		0.90	

Fig. 1. Arching Coefficients and Heger Earth Pressure Distribution

The effects of the weight of the pipe and the fluid in the pipe should also be considered as external loads that cause bending, thrust and shear in the pipe. The supporting reaction for pipe weight is considered to be distributed over an arc of 30 degrees since the pipe is usually placed on the bedding before additional soil is placed in the haunch region below the pipe. The fluid weight is considered to have the same distribution of support reaction as that assumed for earth load for a particular installation type. Live load effects from surface traffic or superimposed surface loads are also given in ASCE 15-93 and added to the other load effects with the same distribution of support reaction as that used for vertical earth load.

Stress Analysis

Stress resultant moments, thrusts and shears caused by external loads and internal pressures are obtained by elastic analysis of the uncracked pipe subject to the loads and load distributions specified in ASCE 15-93 for the selected installation type.

External Load

The analysis for external loads is expedited using the computer program PIPECAR (Version 2.1) [3], and selecting the SIDD Earth Loads (Fig. 1) and Installation Type to obtain the governing moments, thrusts and shears for the selected standard installation. The governing stress resultants for design can also be obtained using non-dimensional coefficients for moments, thrusts and shears at the governing design locations for each of the standard installations multiplied by the respective pipe weight, total earth load, and fluid weight for shear and thrust and by the respective weights times radius to wall centerline for moment. These coefficients are tabulated in the Commentary to the proposed ASCE Standard Practice for Direct Design of Buried Reinforced Concrete Low-Head Pressure Pipe.

Internal Pressure

The axial tensile stress resultants (forces/unit length) in the concrete pipe wall produced by internal operating pressure, p_d, and transient pressure, p_t, are:

Operating Pressure $\qquad\qquad\qquad N_p = -b\ p_d\ D_i/2$ $\qquad\qquad$ (1)

Operating plus transient pressure $\qquad N_{pt} = -b\ (p_d + p_t)\ D_i/2$ $\qquad$ (2)

where $b = 12$ in. (1000 mm) and tensile stress resultants (forces/unit length) are taken as negative.

Design Limits

Because of the successful experience over many years using the pressure limits and flexural design criteria given in ASTM C361 and AWWA C302 and M9, the proposed ASCE Standard for Direct Design of Buried Reinforced Concrete Low-Head Pressure Pipe incorporates the same design limits and tensile stress (strain) limits for concrete and reinforcement. These are summarized as follows:

- Concrete stress limit from internal pressure alone: $\quad f_{ct} = 4.5\ \sqrt{f_c'}$ $\qquad$ (3)

- Reinforcement stress limit from internal operating
  pressure alone: $\qquad\qquad\qquad\qquad\qquad\qquad f_s = 16{,}500 - 75p_d$ $\quad$ (4)

- Reinforcement stress limit from operating+transient pressures alone: $f_s = 16,500$ psi   (5)

- Reinforcement yield stress limit for factored load design: $f_y = 40,000$ psi   (6)

The above concrete stress limit for internal pressure alone governs the maximum internal design pressure that can be provided in a pipe of given diameter, wall thickness and concrete design strength.

Flexural Design for Combined External Load, Fluid Weight and Internal Pressure

Design is based on strength concepts using flexural-axial tensile strength load factors of 1.7, an axial compression load factor of 1.0 and a strength reduction phi factor of 0.95 for load combinations that produce maximum tension. The depth of the compressive stress block under combined bending and compression from external load and axial tension from internal pressure is:

$$a = d \left[1 - \sqrt{1 - 2 \, \frac{M_{ufl} + N_u (d - 0.5h)}{0.85 \, f_c' \, bd^2}} \right] \qquad (7)$$

N_u is + when compression and - when tension.

If the calculated depth of compressive stress block, a, is positive, the required amount of reinforcing on the tension face is calculated as:

$$A_{s1} = \frac{0.85 \, f_c' \, ab - N_u}{\phi_f \, f_y} \qquad (8)$$

Normally, reinforcing is provided near both faces at all cross sections of low-head pressure pipe. However, if "a" is positive, a single line of reinforcing is acceptable. Typically, this arrangement is limited to smaller pipe sizes below about 900 mm (36 in.) diameter.

If the axial tension produced by internal pressure is large relative to the bending produced by external load, the above Equation (7) may show a negative value for the depth of stress block "a". In this case, the combined flexural and axial tension effects produce tension over the entire cross section at that location and a second line of reinforcing must be provided near the face of the pipe wall opposite the face having maximum tension. In this case, the required reinforcing near the face with maximum tension is calculated as:

$$A_{s1} = \frac{M_{ufl} - N_u (d_s + 0.5h - d)}{\phi_f \, f_y \, d_s} \qquad (9)$$

The required minimum reinforcing on the opposite face should be determined using a load factor for bending moment of 1.0 because for this case, bending is providing compression that is reducing the net tension on the opposite face with increase in moment. Thus, for the case where "a" is negative, the minimum tensile reinforcement near this face should be calculated using a load factor of 1.0 on bending moment and axial compression produced by external load and 1.7 on axial tension from internal pressure to determine the minimum required reinforcement area near the opposite face:

$$\min A_{s2} = \frac{-M_{uf2} - N_u (d - 0.5h)}{\phi_f f_y d_s} \qquad (10)$$

The maximum reinforcement design yield strength that is permitted for use in Equations (6), (8), (9), and (10) is 40,000 psi. Considering that the specified load factor and phi factor for tensile reinforcing is 1.7 and 0.95, respectively, these provisions limit the working stress in the tensile reinforcing, based on the assumption of a cracked section, to 22,353 psi, or a strain of about 0.00077. Equations (4) and (5) further limit the reinforcement stress caused by pressure alone and assuming cracked concrete. These requirements greatly limit the concrete strain related to through wall cracking and potential leakage.

Maximum Moment as Limited by Radial Tension and Compression

Two additional types of flexural behavior may limit the amount of external load that can be placed on a reinforced concrete low-head pressure pipe or require radial stirrup reinforcing transverse to the main circumferential reinforcing. These are strength to resist failure by radial tension or by flexural compression. The introduction of internal pressure does not modify these effects so the radial tension and circumferential compression strength criteria previously developed for gravity flow pipe are equally applicable to low-head pressure pipe. Separate calculations for the effects of external load without internal pressure are required to evaluate a low-head pipe design for the above two special flexural effects. A load factor of 1.4, instead of the 1.7 used for flexural tension, is specified in the low-head standard for evaluating radial tension and compression limits. This reduced load factor is used because these limits are not related to leakage, but are similar to the same requirements in ASCE 15-93 which specifies a load factor of 1.3 for radial tension and circumferential compression limits.

Radial Tension

When bending due to external load produces tension in the inner reinforcing at the crown and invert of a reinforced concrete pipe, the curvature of the reinforcing causes radial tensile stresses to be induced in the pipe wall [4]. These must be limited to avoid failure by radial tension (i.e., "slabbing"), or special stirrup reinforcing must be provided to resist these radial effects. A separate calculation of the circumferential reinforcement area needed for the flexural effects of external load only is compared with the maximum circumferential reinforcement area that if stressed to the design yield strength produces radial forces that reach the limiting radial tension concrete strength [4].

The minimum reinforcement area for the tensile flexural effects near the inner face of the pipe wall at the invert, or crown, from external load multiplied by 1.4 for bending and 1.0 for axial compression without internal pressure is:

$$a = d \left[1 - \sqrt{1 - 2 \frac{M_{ur} + N_{ure} (d - 0.5h)}{0.85 f_c' bd^2}} \right] \qquad (11)$$

$$A_{sf} = \frac{0.85\ f_c'\ ab\ -\ N_{ure}}{\phi_f\ f_y} \tag{12}$$

The maximum flexural reinforcement area for external load only, $A_{s\ fmax}$, without stirrups at these locations, as limited by radial tension strength, is:

$$A_{sf\ max} = (\frac{b}{12})\ (16r_s\ \sqrt{f_c'}\ \left(\frac{\phi_r}{\phi_f}\right)\ F_{rt})\ /\ (f_y) \tag{13}$$

where $b = 12$ in.

F_{rt} is an empirically determined size factor and is given as:

for 12 in. $\leq D_i \leq$ 72 in. $\qquad F_{rt} = 1 + 0.00833\ (72 - D_i) \tag{14}$

for 72 in. $< D_i \leq$ 144 in. $\qquad F_{rt} = \dfrac{(144 - D_i)^2}{26,000} + 0.80 \tag{15}$

for $D_i >$ 144 in. $\qquad F_{rt} = 0.8 \tag{16}$

If a particular installation requires greater flexural reinforcement, radial stirrup reinforcing may be provided and must be securely anchored to each line of the inner circumferential reinforcement. Design equations for stirrup reinforcement are given in ASCE 15-93. See also [4].

Compression

The maximum flexural tensile reinforcing for external load with the pipe empty multiplied by 1.4 for both bending and compressive thrust, as limited by concrete compressive strength without compression reinforcement and associated radial ties, is:

$$A_{sf\ max} = \left(\left[\frac{5.5 \times 10^4\ g'\ \phi_f\ d}{(87,000\ +\ f_y)} \right] - 0.75 N_{ure} \right)\ /\ (f_y) \tag{17}$$

$$g' = bf_c' \left[0.85 - 0.05\ \frac{(f_c'\ -\ 4,000)}{1,000} \right] \qquad \begin{array}{l} g'_{max} = 0.85\ bf_c' \\ g'_{min} = 0.65\ bf_c' \end{array} \tag{18}$$

where $b = 12$ in.

Compressive strength is seldom found to be a limiting design criteria in practical reinforced concrete pipe installations although special designs using very deep burial or thin walls are cases where compressive strength could be a governing consideration.

Shear Strength

Design practice for reinforced concrete gravity flow pipe has long recognized the importance of shear or diagonal tension as a governing mode of failure for pipe under high earth covers, or in installations with concentrated support conditions. Design equations given in ASCE 15-93 for shear strength were developed from the evaluation of many 3-edge bearing tests, as well as other tests on beams and frames. However, no pipe test data for test pipe subject to combined external load and internal pressure were available to evaluate the effects of combined bending and axial tension on shear strength. Thus, in ASCE 15, the reduction of shear strength produced by axial tension is taken into account using an empirical factor for direct tension force taken from ACI 318, Building Code and Commentary.

Recent research has shown that this factor gives excessive strength reductions for structures like low-head pressure pipe where tensile strain in the reinforcement is limited as required by the existing low-head pressure pipe standards. The absence of combined load and internal pressure test data and the positive effects achieved relative to shear strength by limiting tensile strains may explain why existing standards for low-head pressure pipe do not contain procedures for calculating shear strength.

New shear design procedures for combined shear and axial tension are included in the recently promulgated AASHTO LRFD Bridge Specification. These procedures are based on extensive research by Vecchio and Collins [5][6] which shows that under combined bending and axial tension, shear strength of reinforced concrete structural members can be controlled by limiting the tensile strain in the reinforcement. The proposed ASCE Standard Practice for Direct Design of Buried Reinforced Concrete Low-Head Pressure Pipe uses the results of this research as implemented in the new AASHTO LRFD Bridge Specification as the basis of the following equations for calculating the shear strength of low-head pressure pipe subject to combined bending from external load and axial tension from internal pressure:

$$V_c = 2.0 \, \phi_v \, bd \, \sqrt{f_c'} \left[\frac{F_d \, F_{ex}}{F_c} \right] \tag{19}$$

where

b = 12 in. and the following modification factors are applied:

$$\text{Size Factor:} \quad F_d = 0.8 + \frac{1.6}{d} \tag{20}$$

max F_d = 1.3, for pipe with two cages or a single elliptical cage
max F_d = 1.4, for pipe through 36 in. diameter with a single circular cage
max d = 16 in.

$$\text{Strain Factor:} \quad F_{ex} = 2.2 - 6\epsilon_x^{0.25} \tag{21}$$

$$0.0 > \epsilon_{xu} > 0.002$$

where:

$$\text{Strain:} \quad \epsilon_{xu} = \frac{M_{uv}/d_v + 0.5V_{uv} \cot \theta_v - 0.4 \, N_{uve} - 0.5 \, N_{up}}{E_s \, A_{si}} \tag{22}$$

Angle of Diagonal Crack: $\theta_v = 37/F_d$, degrees   (23)

Curvature Factor: $F_c = 1 \pm \dfrac{d}{2r}$ (24)

(+) tension on the inside of the pipe
(-) tension on the outside of the pipe

The section where shear is maximum usually is the governing design location for shear strength where the factored shear stress resultant must be less than V_c from Eq. 19 or radial stirrups must be provided. A load factor of 1.4 and a capacity reduction factor $\phi_v = 0.90$ for shear strength are specified in the proposed ASCE Standard Practice for Direct Design of Buried Reinforced Concrete Low-Head Pressure Pipe. See ASCE 15-93 for design of stirrup reinforcing for shear strength. See also [4].

Design Results

Designs using the proposed ASCE Standard Practice for Direct Design of Buried Reinforced Concrete Low-Head Pressure Pipe for four standard diameters of low-head pressure pipe with 20 ft of 100 lb/cu ft earth in both Type 1 and Type 2 SIDD Installations subject to a maximum internal pressure head of 125 ft are presented in Table 1. The design results for each of the Type 1 and Type 2 installations are compared with the tabulated design given in ASTM C361 for each pipe size and this combination of maximum external load and internal pressure.

The design results show that for pipe over the range of diameters and standard wall thicknesses that are given in the table, the designs given in ASTM C361 tables have slightly more reinforcement than required by the SIDD procedures for Type 1 installations and slightly less than required for Type 2 SIDD Installations. This is not surprising since the SIDD designs are governed by combined flexure and axial tension and the criteria used for flexure and tension are essentially the same in SIDD and ASTM C361.

A significant finding from these design examples is that the C361 pipe designs have adequate shear and radial tension strength (except for the 9.5 in. wall, 108 in. diameter, SIDD Type 2 design, which has low radial tension strength). Since there are no criteria for determining shear or radial tension strength of reinforced concrete pressure pipe given in the existing standards for design of low-head pressure pipe, it was important to check these criteria for the tabulated designs of pipe having the deepest burial combined with the highest pressure.

Table 1.   Summary of Example SIDD Low-Head Pressure Pipe Design Results and Comparison with ASTM C361 Table Design

Pipe Internal Diameters:	42, in., 60 in., 84., 108 in.
Pipe Wall Thicknesses:	4.5 in., 6 in., 8 in., 9.5 in.
Reinforcement Arrangement:	2 circular cages
Concrete Clear Cover:	1 in.
Design Concrete Compressive Strength, f_c':	4500 psi
Earth Cover Over Pipe:	20 ft
Unit Weight of Earth:	100 lb/cu ft
Pressure Head:	125 ft (54 psi)
Minimum Yield Strength of Reinforcement:	40,000 psi

Pipe Inside Diam. (in.)	Wall Thickness (in.)	Installation Type (SIDD or C361)	Reinforcement Areas		Radial* Tension Index	Shear* (Diagonal Tension) Index
			Inner at Invert (in.²/ft)	Outer at Springline (in.²/ft)		
42	4.5	1	0.80	0.63	0.44	0.48
		2	0.96	0.70	0.63	0.72
		C361	0.86	0.58	–	–
60	6.0	1	1.12	0.88	0.54	0.58
		2	1.33	0.96	0.75	0.84
		C361	1.22	0.83	–	–
84	8.0	1	1.59	1.22	0.68	0.68
		2	1.89	1.34	0.94	0.98
		C361	1.79	1.22	–	–
108	9.5	1	2.17	1.64	0.86	0.79
		2	2.59	1.80	1.16**	1.12
		C361	2.32	1.60	–	–

*   Note: Radial tension and shear strength indexes are the ratio of 1.4 times service load design condition to calculated maximum strength.

** Requires stirrup reinforcing in invert region.

Conclusions

The proposed new ASCE Standard Practice for Direct Design of Buried Reinforced Concrete Low-Head Pressure Pipe in Standard Installations (SIDD) provides a complete design procedure for this important type of pipe. The design criteria and ultimate strength methods for combined flexure produced by external loads and axial tension produced by internal pressure are essentially the same as those used for many years by designers complying with ASTM C361, or AWWA C302 and Manual M9. However, the proposed new standard installations are the more quantitative and rational installations termed the SIDD standard installations and defined in ASCE 15-93.

The design provisions also include checks for the important flexural design limit resulting from the radial tension strength of the concrete that maintains the curvature of the inner reinforcing at the crown and invert of the pipe and the significant shear (diagonal tension) strength limit. These limits frequently govern the design of pipe under high fills or with concentrated support conditions from poor bedding. Defining design criteria for these limits represents a significant advance in the state of the art of reinforced concrete low-head pressure pipe design.

The radial tension limit is unaffected by internal pressure and thus is the same as that given in ASCE 15-93 for gravity flow pipe. The shear strength limit is determined using a new design procedure derived from the recently promulgated AASHTO LRFD Bridge specification and extensive recent research on flexural members subject to combined bending, axial tension and shear. This information has been used because test data on pipe subject to combined shear, bending and axial tension has not been found and the empirical method used in ACI 318 has been found to be so conservative that successful experience with previous designs of low-head pressure head pipe would be contradicted. The validity of the proposed shear strength design procedures is demonstrated by the successful design of reinforced concrete pressure pipe based on strain limits, similar to the new approach in the AASHTO LRFD Bridge Specification.

Methods given in ASCE 15-93 for designing radial stirrup reinforcing may be used to increase the radial tension and/or the shear strength of low-head pressure pipe that require resistance to these types of structural effects to allow adequate structural safety.

References

[1]   Heger, F.J., "New Installation Designs for Buried Concrete Pipe," *ASCE Pipeline Infrastructure Proceedings*, 1988.

[2]   *Concrete Pipe Design Manual*, American Concrete Pipe Association, Vienna, VA, Seventh Printing, 1987.

[3]   McGrath, T.J., Tigue, D.B., Rund, R.E., and Heger, T.G., "PIPECAR User and Programmer Manual, Version 2.1," U.S. Department of Transportation, Federal Highway Administration, July 1994. (Available from McTrans, University of Florida, 512 Weil Hall, Gainesville, FL 32611-2083.)

[4]   Heger, F.J., *Concrete Pipe Technology Handbook*, reviewed, edited and published by American Concrete Pipe Association, Vienna, VA, March 1993.

[5]   Vecchio, F.J., and Collins, M.P., "Predicting the Response of Reinforced Concrete Beams Subjected to Shear Using the Modified Compression Field Theory," *Structural Journal*, Vol. 85, No. 4, American Concrete Institute, May-June 1988.

[6]   Vecchio, F.J., and Collins, M.P., "The Modified Compression Field Theory for Reinforced Concrete Elements Subjected to Shear," *Journal of the American Concrete Institute*, Vol. 83, No. 2, March/April 1986.

Gerald R. Frederick[1] and Kassim M. Tarhini[2]

Structural Evaluation of Three-Sided Concrete Culverts

Reference: Frederick, G. R. and Tarhini, K. M., "Structural Evaluation of Three-Sided Concrete Culverts," *Concrete Pipe for the New Millennium, ASTM STP 1368,* I. I. Kaspar and J. I. Enyart, Eds., American Society for Testing and Materials, West Conshohocken, PA, 2000.

Abstract: Three-sided concrete culverts can be used to replace short span bridges and multiple sections (barrels) of four-sided concrete box culverts. ASTM Standards do not specify designs for four-sided concrete box culverts with span lengths exceeding 12 ft (3.6 m) nor do they discuss the three-sided concrete culverts. This paper describes the analysis and design of three-sided flat-top precast reinforced concrete culverts with span length ranging between 14 ft (4 m) and 36 ft (11 m). It was shown that the AASHTO distribution width used in performing plane frame analysis gave similar results to the three-dimensional finite element analysis. Therefore, ten shallow structures were selected for analysis and design. The culverts were subjected to live load plus impact, dead load, and lateral earth pressure. The slab and wall thicknesses were selected so that no shear reinforcement is needed. The ultimate strength design was used to determine main reinforcing steel. AASHTO and ASTM recommendations were used to determine the reinforcing steel in transverse direction.

Keywords: Analysis, design, three-sided culverts, precast

The Federal Highway Administration (FHWA) reports about 30% of the nation's 589 243 bridges are deficient functionally or structurally. The majority of the structurally deficient bridges are short spans, averaging less than 50 ft (15 m) in length. This bridge inventory accounts for structures with span lengths greater than 20 ft (6 m). Prefabricated concrete culverts are often economical alternatives for replacing deteriorating short span bridges and cast in place culverts. Currently, design standards exist for two categories of four-sided box culverts: ASTM Specification for Precast Concrete Box Sections for Culverts, Storm Drains and Sewers with less than 2 ft (0.6 m) of Cover Subject to Highway Loadings (C 850) establishes the designs of 42 box sections and ASTM Specification for Precast Concrete Box Sections for Culverts, Storm Drains and Sewers (C 789) establishes standard designs for another 42 box sections with more than 2 ft (0.6 m) of soil cover. The maximum span length of a standard precast four-sided concrete box

[1] Professor of Civil Engineering, University of Nevada, Las Vegas, Las Vegas, NV 89154-4015
[2] Associate Professor of Civil Engineering, Valparaiso University, Valparaiso, IN 46383

culvert is 12 ft (3.6m). This span length is sometimes too small to handle heavy water flow which may require the use of multiple sections placed side by side. In this case, the walls of adjacent culverts will act as a pier which may obstruct the flow and be associated with flooding problems. Therefore, developing new four-sided box sections with longer spans could prove to be an economical alternative to multiple sections. However, as the span length increases, the weight of the section will also increase (unless laying length is decreased) and may cause transportation problems associated with moving the box sections to the job site.

Three-sided concrete culverts (rigid frame structures) have been developed to replace short span bridges ranging between 14 ft (4 m) and 36 ft (11 m) and multiple sections (barrels) of four-sided concrete box culverts. Three-sided concrete culverts may have either a flat-top or an arch top. Additionally, the flat-top concrete culverts may be reinforced or prestressed. In this paper, three-sided flat-top precast concrete culverts utilizing ordinary reinforcing steel will be discussed. A typical three-sided flat-top concrete culvert is shown in (Figure 1). Such a structure is commonly supported on strip footings; accordingly, the waterway usually has a natural bottom. ASTM Standards do not specify designs for four-sided concrete box culverts with span lengths exceeding 12 ft (3.6 m), nor do they discuss the analysis and design of three-sided concrete culverts. Therefore, this paper will discuss the analysis and design procedures of three-sided flat-top concrete culverts with less than 2 ft (0.6 m) of soil cover and then present a design summary of ten structures.

Structural Analysis

The geometry of a three-sided concrete culvert can be modeled using either solid brick elements or shell elements to perform three-dimensional finite element analysis (FEA). Frederick et al. [1 and 2] reported detailed finite element analyses on several flat-top three-sided culverts. The finite element results were compared with rigid frame analysis subject to modified design live loading. The finite element analysis program, SAP90, was used to analyze the structural behavior of these culverts. The concrete walls and slab were idealized using quadrilateral shell elements with six degrees of freedom at each node. AASHTO HS20 wheel loads, including impact, of 20.8 Kips (92.6 KN) were positioned at various nodes or distributed over a tire-print area at mid-span of the top slab. All of the flat-top three-sided culverts analyzed using FEA did not consider (nor rely upon) the lateral earth pressure effects for stability or strength. The culverts were assumed to safely support design wheel loads as free-standing units.

Two mesh sizes that were investigated using shell elements 1 ft x 1 ft (0.3 m x 0.3 m) and 0.5 ft x 0.5 ft (0.15 m x 0.15 m) yielded similar bending moment and deflection results. Several wheel load positions (placed one at a time) on the top slab were investigated using 0.5 ft x 0.5 ft (0.15 m x 0.15 m) rectangular shell elements. The maximum value of the bending moment, deflection, and bending stresses in the top slab decreased as the concentrated wheel load moved from the edge toward the center (in 1 ft increments) of the top slab at mid-span. Applying the wheel load over a tire print area reduced the stress concentration effects. It was shown that when the wheel load was distributed over the tire print area, the results were similar to wheel loads placed at nodes

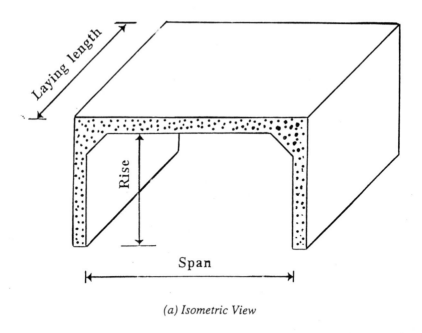

(a) Isometric View

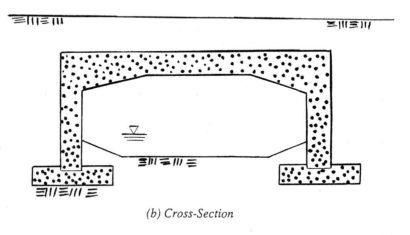

(b) Cross-Section

Figure 1 - *Typical Three-Sided Precast Concrete Culvert*

2 ft (0.6 m) or 3 ft (0.9 m) from the edge. The finite element results due to load near the center or applied over a tire print area compared well with the rigid frame analysis. Furthermore, Tarhini et al. [3] reported favorable comparison between the finite element results and experimental data obtained by testing several one-sixth size models of two culverts. The experimental data were similar to the results obtained using rigid frame analysis.

The typical analysis of a precast concrete box culvert is to treat the three-dimensional structure as a series of slices that behave as unit width rigid plane frames. The corresponding live load is determined using the AASHTO [4] wheel load distribution width, E, for slabs with main reinforcement parallel to traffic:

$$E = (4.0 + 0.06S) \leq 7 \text{ ft} \qquad (1)$$

where

S = effective span length in feet
(or E = 1.2 + 0.06 S where S is the effective span length in meters).

Here, the effective span length was interpreted to be the clear span minus the length of one haunch. Note that this is the AASHTO recommendation when 45 degree haunches are present; here it was used regardless of the haunch angle.

Structural Design

The three-sided flat-top concrete culverts were analyzed and designed as rigid frame structures. All the structures considered in this paper had a haunch 1 ft (0.3 m) vertically by 2 ft (0.6 m) horizontally at its two upper corners. The lower ends of the side walls rest upon strip footings and are assumed to be hinged. The frame members are considered to have varying moment of inertia due to the haunches between the walls and top slab. Hence, if a classical method of structural analysis is used, some coefficients or procedures require modification since the members are not prismatic. Alternately, a structural analysis computer program could be used to model the structure (plane frame or 3D FEA) and perform the analysis.

The load cases to be considered in the analysis and design of three-sided flat-top culverts are:

Dead load of material above the culvert
Dead load of the culvert
Live load and impact
Lateral earth pressure

These load cases are shown schematically in (Figure 2). As in any precast concrete structure, the handling and transportation stresses must be considered since they may control the design. The three-sided flat-top culverts considered in this paper were assumed to have a 5 in (125 mm) thick pavement placed directly upon the top slab. The soil adjacent to the structure was assumed to be granular with a unit weight of 120 pcf

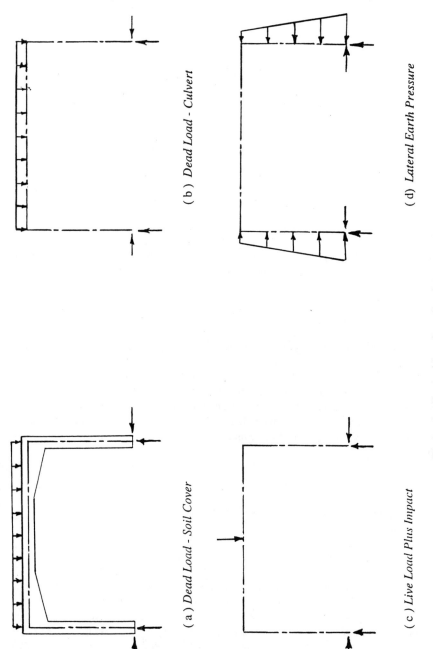

(a) Dead Load - Soil Cover

(b) Dead Load - Culvert

(c) Live Load Plus Impact

(d) Lateral Earth Pressure

Figure 2 - Typical Loading Conditions on the Culvert

(1920 kg/m^3) and an angle of internal friction of 30°. Active earth pressure was assumed to act on the sides of the structure. Stream flow pressure was not included since it tends to reduce the design forces at critical sections. The design live load was selected to be the AASHTO HS20 wheel load of 16 Kip (71.2 KN) plus 30% impact. The concrete strength was selected to be 5 Ksi (34.4 MPa) and the yield strength of the welded wire fabric was taken to be 65 Ksi (448 MPa). Two inches of concrete cover was assumed for all reinforcing steel.

The maximum bending moments and shear forces are used to verify the estimated wall and slab thicknesses, to determine the required reinforcing steel, to verify the shear strength, to check the crack control provisions and to check the live load deflection criteria. Distribution reinforcement for slabs with main reinforcement parallel to traffic, as specified in AASHTO, must also be provided. Following the ASTM practice for C 850 box culverts, the distribution reinforcing, as a minimum, was taken to be shrinkage and temperature reinforcing. The required amounts of shrinkage and temperature reinforcing are placed near both surfaces of the slab. Hence, twice the required amount of shrinkage and temperature reinforcing is provided. In general, the thickness of the various slabs have been selected so that shear reinforcing is not required. Usually, the calculated ultimate shear stress is less than 65 % of its permissible value. Design summaries for various combinations of spans and rises of ten three-sided flat-top concrete culverts are presented in (Table 1). The reinforcing designations are shown in (Figure 3) for a typical three-sided flat-top culvert.

The analysis and design procedures used in developing (Table 1) was further confirmed by performing a laboratory test on a one-sixth size scale model of a 20 ft (6 m) span by 10 ft (3 m) rise three-sided flat-top concrete culvert. At the scaled design load of 580 lbs (2.58 KN) (representing a 16 Kip wheel load plus 30% impact), the maximum observed compressive stress in the concrete was 760 psi (5.24 MPa) with a maximum vertical deflection of 0.028 inches (0.71 mm). In a full size structure, the maximum compressive stress in the concrete is expected to be 760 psi (5.24 MPa) with a maximum vertical deflection of 0.17 inches (4.3 mm). This test was performed without loading or support applied to the walls. The ultimate load sustained by the one-sixth size model was 2720 lbs (12.1 KN) or 4.8 times the scaled design load.

Three-sided flat-top culverts are generally fabricated with (shear) keyways between adjacent units. Shear keys are then formed by filling the keyways with grout during erection. Additionally, prestressed tie-rods may be placed through adjacent units to enhance the effectiveness of the shear keys (as is commonly done with precast box beams). The behavior of shear keys is not reported in this investigation.

Conclusion

This paper has demonstrated that three-sided flat-top concrete culverts can be satisfactorily analyzed and designed using plane rigid frame or three-dimensional finite element analysis. The advantage of 3D FEA is that it provides values for the transverse

Table 1 – *Three-sided Flat-top Concrete Culvert Design Summary*

Span x Rise (ft x ft)	Thickness (in)		Reinforcing Steel Areas (Sq. in./ft.)							
	Top Slab	Wall	A_{S1}	A_{S2}	A_{S3}	A_{S4}	A_{S5}	A_{S6}	A_{S7}	A_{S8}
14 x5	14	12	0.34	0.34	0.34	0.57	0.51	0.29	0.29	0.29
14 x 9	14	12	0.34	0.34	0.34	0.67	0.48	0.29	0.29	0.29
16 x 5	15	12	0.36	0.36	0.36	0.61	0.59	0.29	0.29	0.29
16 x 9	15	12	0.36	0.36	0.36	0.73	0.54	0.29	0.29	0.29
24 x 5	20	15	0.57	0.48	0.48	0.74	0.74	0.36	0.36	0.36
24 x 9	20	15	0.48	0.48	0.48	0.85	0.73	0.36	0.36	0.36
28 x 5	24	17	0.63	0.58	0.58	0.87	0.80	0.41	0.41	0.41
28 x 9	24	17	0.58	0.58	0.58	0.90	0.76	0.41	0.41	0.41
32 x 5	28	19	0.67	0.67	0.67	1.03	0.85	0.46	0.46	0.46
36 x 8	30	20	0.72	0.72	0.72	1.08	1.01	0.48	0.48	0.48

Note: 1 ft. = 0.3048 m and 1 in^2 = 645 mm^2

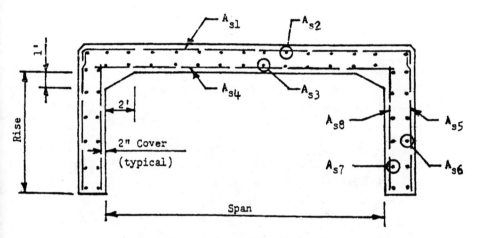

Figure 3 - *Reinforcing Designations*

bending moments and shear forces. However, these bending moments are low enough so that, for the cases investigated, the AASHTO and ASTM requirement for distribution reinforcing provide adequate strength. Therefore, for a given shallow three-sided flat-top concrete culvert, the engineer can analyze the structure using plane frame analysis subjected to various combinations of dead loads, live load plus impact, and lateral earth pressure.

References

[1]   Frederick, G.R., Tarhini, K.M., and Mabsout, M.E., "Development of a Design Procedure for Three-Sided Concrete Culverts," *Structural Performance of Pipes*, G.F. Mitchell, S.M. Sargand, and K. White, Eds., Ohio University, Athens, OH, 1998, pp.63-73.

[2]   Frederick, G.R., Tarhini, K.M., Mabsout, M.E., and Faraj, C., "Analysis of Three-Sided Precast Concrete Culverts," Transportation Research Board, 76th Annual Meeting, Washington, DC, 1997.

[3]   Tarhini, K.M., Frederick, G. R., Mabsout, M. E., and Faraj, C., "Experimental and Analytical Evaluation of Three-Sided Concrete Box Culverts," *Post Conference Proceedings of the 1997 SEM Spring Conference on Experimental Mechanics,* Bellevue, WA, 1997, pp, 209-213.

[4]   AASHTO, *Standard Specifications for Highway Bridges*, 16th Ed., American Association of State Highway and Transportation Officials, Washington, DC, 1996.

Ronald E. Rund[1] and Timothy J. McGrath[1]

Comparison of AASHTO Standard and LRFD Code Provisions for Buried
Concrete Box Culverts

Reference: Rund, R. E. and McGrath, T. J., **"Comparison of AASHTO Standard
and LRFD Code Provisions for Buried Concrete Box Culverts,"** *Concrete Pipe
for the New Millennium, ASTM STP 1368,* I. I. Kaspar and J. I. Enyart, Eds.,
American Society for Testing and Materials, West Conshohocken, PA, 2000.

Abstract: AASHTO developed the LRFD Bridge Design Specification, with the
intent of replacing the Standard Specifications for Highway Bridges with a reliability
based code that provides more uniform safety for all elements of bridges. Although
many provisions in these two codes are the same, there are important differences that
can have a significant effect on the amount of reinforcement required for buried
precast reinforced concrete box culverts under some conditions.
     AASHTO LRFD Code provisions that differ from the Standard Specifications
include load factors, load modifiers, load combinations that produce the maximum
force effect, multiple presence factors, design vehicle loads, distribution of live load
to slabs and through earth fill, dynamic load allowance, live load surcharge, and
reinforced concrete design methodology for fatigue, shear strength, and control of
cracking. These provisions are compared with the equivalent provisions from the
AASHTO Standard code, and discussed in detail.
     Several combinations of culvert sizes and fill heights are evaluated using pro-
visions from both the Standard and LRFD Specifications. A comparison of results
show that in general, LRFD provisions result in increased design loads and reinforcement areas.

Keywords: reinforced concrete design, box section, culvert, AASHTO, LRFD

Introduction

     In 1994, after about 10 years of development, AASHTO introduced a new and
alternate bridge design specification based on a load and resistance factor design
method titled "LRFD Bridge Design Specification" [1]. Although this specification
is currently allowed as an alternate to the AASHTO Standard Specification for Highway Bridges [2], AASHTO intends to eventually discontinue the Standard Specifications. A comparison of provisions related to the design of precast concrete box
culverts for these two codes show that while many requirements are the same, there
are significant differences that have an impact on design. These differences relate to
load factors, load modifiers, load combinations that produce the maximum force
effect, multiple presence factors design vehicle loads, distribution of live load to
concrete slabs and through earth fill, dynamic load allowance, live load surcharge,

[1] Senior Staff Engineer and Principal, respectively, Simpson Gumpertz & Heger Inc.,
     Consulting Engineers, 297 Broadway, Arlington, MA 02474

and reinforced concrete methodology for fatigue, shear resistance and control of cracking. The following discussions are based on the 1994 AASHTO LRFD Bridge Design Specification with 1996 and 1997 Interims (LRFD) and Standard Specifications for Highway Bridges, Sixteenth Edition, 1996 with 1997 Interims (Standard).

LRFD Approach

Load factor design for concrete structures in Standard is similar to load resistance and factor design requirements in LRFD in that both use factored loads, strength reduction factors and checks on serviceability. However, LRFD expands on these requirements and accounts for variability in predicting loads and resistance of structural elements based on a probabilistic determination of reliability. The goal of the LRFD approach is to provide a more rational design basis with more uniform reliability.

LRFD introduces the limit states concept which dictates that all components meet the following requirement for each limit state:

$$\sum \eta_i \, \gamma_i \, Q_i \le \phi \, R_n \tag{1}$$

Where η_i = load modifier is related to ductility, redundancy and operational importance
   γ_i = load factor
   Q_i = specified load
   R_n = nominal resistance
   ϕ = resistance factor, identical to Standard's strength reduction factor

LRFD defines the following four limit states:

- Service Limit State, which imposes requirements on deflection and crack width under service load conditions;
- Fatigue Limit State, which limits the stress range in reinforcement from application of a single design truck under service load conditions;
- Strength Limit State, which requires that the strength and stability of the structure be adequate for specified load combinations for the design life of the structure; and
- Extreme Limit State, which requires survival of the structure during an event such as an earthquake. LRFD suggests the design of buried culverts for seismic forces need only be investigated when a culvert crosses an active fault. This agrees with findings from condition assessments of buried concrete culverts having experienced significant seismic events [3]. The findings show overall culvert survival but localized damage at features such as head walls, penetrations, and terminations.

Differences Between LRFD and Standard

Load Factors and Load Modifiers

Both the LRFD service limit state and Standard strength design check on serviceability use load factors equal to 1.0. However, for permanent loads under the strength limit state, LRFD has introduced the concept of a range of load factors, each of which can be increased or decreased by a load modifier. A design value load modifier η_i is computed from individual values as:

$$\eta_i = \eta_D \, \eta_R \, \eta_I \geq 0.95 \text{ for maximum value load factors} \tag{2}$$

$$\eta_i = \frac{1}{\eta_D \, \eta_R \, \eta_I} \leq 1.00 \text{ for minimum value load factors} \tag{3}$$

where, η_D relates to ductility and is generally equal to 1.00 for culverts detailed in accordance with LRFD requirements; η_R is associated with the redundancy of the structure (Buried culverts under earth loading are categorized as non-redundant with η_R equal to 1.05, while all other loads are considered redundant and have factors equal to 1.00.); and η_I relates to importance and is determined on the basis of an assessment of the need for continued function and safety.

The most dramatic difference in load factors occurs for live load, where LRFD provisions require $\gamma_i = 1.75$ and Standard requires $\gamma_i = 2.17$ (The effect of this is negated by the introduction of a multiple presence factor of 1.2 for single lane loading, as discussed later). A list of load factors for both LRFD and Standard is given in Table 1.

Table 1 – *Load Factors*

Load	LRFD Load Factors	Standard Load Factors
Self weight	0.90 and 1.25	1.3
Wearing surface	0.65 and 1.50	1.3
Horizontal earth pressure	0.90 and 1.35	0.65 and 1.30
Vertical earth pressure	1.30	1.3
Live load	1.75	2.17
Water	1.00	1.3
Live load surcharge	1.75	2.17
Downdrag	0.45 and 1.80	No Requirement
Construction loads	≥ 1.5	No Requirement

Load Combinations

In general, design loads utilizing maximum load factors are combined to produce the maximum force effect. However, when one design load decreases the effect of another, the minimum load factor is used for the load that decreases the force effect. Load combinations and load factors, as adjusted by load modifiers, for a typical box culvert designed in accordance with LRFD requirements are presented in Table 2.

In Table 2, Load Combination 1 represents the maximum vertical load applied to the culvert in combination with minimum lateral load, Load Combination 2 represents minimum vertical load in combination with maximum lateral load and Load Combination 3 represents maximum vertical load and maximum lateral load.

Design Vehicle Load

LRFD provisions offer two types of design vehicle loads, a design truck or a design tandem, each to be used in combination with a design lane load. The design truck consists of the same wheel loads, and axle spacings as the HS20 truck in Standard; however, the design tandem axle load is 25,000 lb as compared to 24,000 lb for Standard Specification alternate military loading (sometimes referred to as the "Interstate loading"). In addition, LRFD requires the design truck or design tandem load to be combined with a design lane load of 640 lb/ft uniformly distributed over a 10 ft design lane width.

Table 2 – *LRFD Load Combinations and Adjusted Load Factors*

LRFD Load Comb.	Load Factor x Load Modifier							
	Limit State	Self Wgt.	Wearing Surface	Horiz. Earth Load	Vert. Earth Load	Vert. Live Load	Water	Live Load Surch.
1	Service	1.0	1.0	0.5	1.0	1.0	1.0	–
	Fatigue *	–	–	–	–	0.75	–	–
	Strength	1.25	1.50	0.64	1.37	1.75	1.0	–
2	Service	1.0	1.0	1.0	1.0	–	–	1.0
	Fatigue *	–	–	–	–	–	–	0.75
	Strength	0.9	0.65	1.42	1.37	–	–	1.75
3	Service	1.0	1.0	1.0	1.0	1.0	1.0	1.0
	Fatigue *	–	–	–	–	0.75	–	0.75
	Strength	1.25	1.50	1.42	1.37	1.75	1.0	1.75

* LRFD Specifications state that fatigue need not be investigated for buried concrete structures. Since box sections with less than 2 ft of fill are designed with live load applied directly to the top slab, these sections should not be considered buried and are evaluated for fatigue.

Multiple Presence Factor

For spans less than or equal to 15 ft, LRFD provisions require that force effects from design vehicle loading be multiplied by a multiple presence factor, which is dependent on the number of loaded traffic lanes. This factor is similar in concept to the reduction in load intensity provisions from Standard. LRFD specifies a multiple presence factor of 1.2 for one loaded lane, 1.0 for two loaded lanes, 0.85 for three loaded lanes and 0.65 for four or more loaded lanes. In comparison, Standard requires similar factors equal to 1.0 for one or two loaded lanes, 0.90 for three loaded lanes and 0.75 for four or more loaded lanes. For shallow depths of fill, the LRFD multiple presence factor causes a single loaded lane to always produce the maximum force effect. The increased multiple presence factor balances the effect of the reduced live load factor in LRFD noted in Table 1.

Dynamic Load Allowance

LRFD provisions for dynamic load allowance, which account for impact from moving vehicles, are greater than those of Standard. For LRFD, the allowance varies linearly from 33 percent at 0 ft of fill (although the current edition of LRFD indicates 40 percent, this has been changed but is not yet in print) to 0 percent at 8 ft and is not applicable to the design lane load. Standard provisions decrease in ten percent steps from 1.3 at 0 ft of cover to 1.0 at 3 ft or greater cover. In general, LRFD requirements produce significant increases in live load relative to Standard, particularly at depths of 2 ft to 4 ft.

Tire Contact Area

LRFD provisions for tire contact area assume a tire pressure of 125 psi and account for increased length of the tire footprint from the effects of the dynamic load allowance and load factor. Standard provisions base the contact area on a tire pressure of 100 psi but do not consider any increase in contact area from impact or load factor. The overall result of these changes is that LRFD allows a larger tire footprint than does Standard. More significant than the size of the footprint are the provisions in LRFD for including the footprint distribution when applying loads to culverts, as discussed in the following sections.

Distribution of Wheel Loads For Depths of Fill Less Than 2 Feet

For fill depths less than 2 ft, the strip widths (the effective width of slab that resists the applied vehicle load) in LRFD are designated for an axle load, which includes two wheels, while in Standard, strip widths apply to a single wheel. In the direction of travel, LRFD permits an axle load to be distributed over a distance equal to the tire length, whereas Standard does not allow this distribution. Neither code allows increased distribution of live load with increasing fill heights for depths less than 2 ft.

A comparison of equivalent axle strip widths from each Specification (Table 3) shows that LRFD strip widths are smaller for shorter spans with decreasing difference as the span increases. For smaller span culverts designed in accordance with LRFD, application of the narrow widths will result in increased reinforcement requirements. This appears to contradict the 20-plus years of satisfactory performance of culverts furnished under AASHTO M 273 and studies of the behavior of culverts designed in accordance with M 273, by James [4] and Frederick, et al. [5] which indicate that live load forces obtained from Standard strip widths are conservative.

Although not included as a code requirement, the distribution width attributed to one wheel loading a precast box section under 0 to 2 ft of cover should not exceed the section length of a single segment of box section, unless shear connectors or other means are provided for transfer of wheel loads across joints between adjacent box sections.

Distribution of Wheel Loads For Depths of Fill Greater or Equal to 2 Feet

LRFD provisions for distribution of wheel loads through fill often yield greater design forces from wheel loads than Standard, especially at shallow covers. Wheel loads based on LRFD requirements are distributed through fill over an area equal to the tire footprint, with the footprint dimensions increased by either 1.15 times the depth of fill for select granular backfill, or 1.0 for other types of backfill. In contrast, Standard considers a wheel load as a point load and distributes it over a square equal to 1.75 times the depth of fill, regardless of the type of backfill.

Live Load Surcharge (Lateral Live Load Effects)

Live load surcharge effects are typically created when a vehicle approaches a buried culvert and creates additional horizontal earth pressures on the sides of the culvert. In general, LRFD produces greater live load surcharge pressures than Standard for depths of fill of 5 ft or less and less pressure for greater depths. In addition, live load surcharge pressures from AASHTO M 259 [6] and M 273 [7] are much greater than those from LRFD for depths of fill from 0 to 1 ft, and less than LRFD for greater fill heights. In spite of the significant differences in live load surcharge pressures, their impact on reinforcement areas is relatively minor.

Table 3 – *Equivalent Axle Strip Widths (ft)*

Span (ft)	LRFD		Standard (4)
	Positive Moment	Negative Moment	
4 (1, 2)	4.55	5.08	8.44
8 (1, 2)	6.93	6.17	8.88
12 (1, 2)	9.32	7.25	9.32
16 (3)	9.94	9.94	9.76
20 (3)	11.0	11.0	10.20

Notes:
(1) Positive moment axle strip width = $(26 + 6.6*S)/12 \leq 12$ ft, where S = mean span (ft), single loaded lane
(2) Negative moment axle strip width = $(48.0 + 3.0*S)/12 \leq 12$ ft
(3) For spans greater than 15 ft, positive and negative moment, Axle strip width = $(10.0 + 5.0* (L_1*W_1)**0.5)*1.2$, where L_1 = clear span (ft), W_1 = 20 ft bridge width A 1.2 factor is included in the expression for strip width to eliminate the multiple presence factor (1.2) that is incorporated into the basic expression. LRFD suggests a multiple presence factor of 1.0 for spans greater than 15 ft.
(4) Equivalent axle strip width = $(2)(4 + 0.06*S)$, where S = span - haunch dimension (ft)

Combined Effect of Live Load Requirements for LRFD and Standard

To assess the combined effect of requirements relating to live load, including load factors, load modifiers, multiple presence factor, dynamic load allowance, tire contact area, live load surcharge, and distribution of wheel loads to reinforced concrete slabs and through fill, a comparison has been made of maximum factored positive moment and negative moment at the tip of the haunch in the top slab of a 4 ft span by 4 ft rise, and 8 ft span by 8 ft rise culvert at depths of fill ranging from 0 ft to 8 ft. Results are presented in Figures 1 and 2 and show that factored positive and negative moments from LRFD are greater than Standard at most depths of fill with the effect most pronounced at 0 ft for positive moment.

Vertical and Horizontal Earth Pressures

Requirements for determining vertical earth pressures are identical for LRFD and Standard. Both specifications use soil-structure interaction factors that are applied only to vertical earth loads, vary with depth of fill, and are always greater than or equal to 1.0. For horizontal earth pressures, both codes allow the use of the equivalent fluid method when more accurate information is not available. LRFD designates a minimum horizontal pressure based on an equivalent fluid unit weight of soil of 0.030 kcf and provide typical equivalent fluid unit weights for several types of soils including 0.045 kcf to 0.055 kcf for dense to loose sands or gravel, 0.060 kcf for compacted silt, 0.070 kcf for compacted lean clay, and 0.080 kcf for compacted fat clay. For installations where compacted granular fill is placed along the sides of the culvert, the LRFD range of equivalent fluid unit weights of soil is similar to the Standard specified range of 0.030 kcf to 0.060 kcf. Where a reduction in horizontal earth pressure increases the effects from other loads, such as increased positive moment in top and bottom slabs of box culverts, LRFD and Standard require that reduced horizontal earth pressures be considered. Where precise information is not available, a 50% reduction to pressure is suggested. This reduced pressure need only be used in combination with the maximum load factors presented in Table 1.

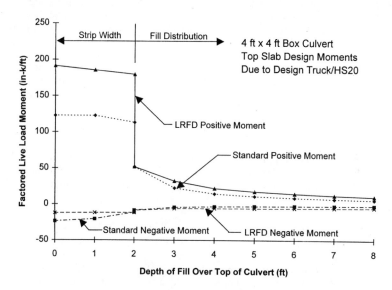

Figure 1 – *Live Load Moments for 4 ft x 4 ft Box Culverts*

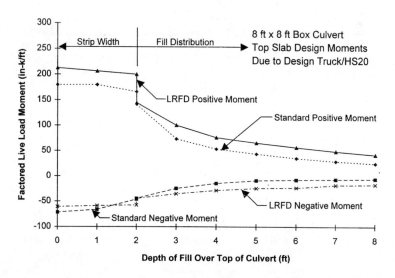

Figure 2 – *Live Load Moments for 8 ft x 8 ft Box Culverts*

Shear Resistance of Culverts without Shear Reinforcement for Fill Heights Greater or Equal to 2 Feet

For single cell box culverts under fill heights of at least 2 ft, the requirements for minimum shear resistance of slabs without shear reinforcement, in both codes are the same, allowing an ultimate shear stress of $3\sqrt{f_c'}$ (f_c' is the design compressive strength of concrete in psi). More detailed provisions based on moment to shear and reinforcement ratios are also identical. Even though coefficients in the equations used to determine shear resistance appear different, the difference is due to units for f_c' taken as ksi for LRFD and psi for Standard.

Neither code mentions the applicability of this shear requirement to sidewalls; however, since this requirement is appropriate for members subject to uniformly distributed loads, we assume the intent of both codes is to apply this allowable shear resistance to both slabs and sidewalls.

Shear Resistance of Culverts without Shear Reinforcement for Fill Heights Less Than 2 Feet

LRFD incorporates a new approach for evaluating shear resistance in culverts with less than 2 ft of cover that is based on the modified compression-field theory. The following expressions are used to determine the nominal shear resistance of members without shear reinforcement:

LRFD	Standard

$$V_c = 0.0316\ \beta\sqrt{f_c'}\ b_v\ d_v\ (f_c'\ \text{in ksi}) \quad (4) \qquad V_c = 2\sqrt{f_c'}\ b\ d\ (f_c'\ \text{in psi}) \quad (5)$$

where b_v = b = effective width of the section, usually equal to 12 (in.)
 d = d_e = depth from the compression face to the centroid of the tension reinforcement (in.)
 d_v = the effective shear depth (in.). Note that d_v is equal to the distance between the resultants of the tensile and compressive forces, as measured perpendicular to the neutral axis of the member, and does not need to be less than 0.9 times d_e or 0.72 times the total thickness of the member.
 β = factor which accounts for the ability of diagonally cracked concrete to transmit tension
 V_c = the nominal shear resistance provided by tensile stresses in the concrete (k)

For member thicknesses less than 16 in., LRFD offers a simplified procedure for evaluating the above expression and a general procedure which is much more complex. The general procedure requires calculation of the shear stress, strain in the reinforcement, and θ, which is defined as the angle between the diagonal compressive stresses and longitudinal axis of the member. Results can only be obtained by an iterative solution. This complex procedure is not within the scope of this paper but is discussed in detail by LRFD and by Collins and Mitchell [8].

For nominal shear resistance of members without shear reinforcement, the simplified procedure given in LRFD assumes β equal to 2.0. Thus, if d_v is assumed to be equal to 0.9 times d_e, LRFD nominal shear resistance is 90 percent that of Standard.

Shear Resistance of Culverts with Shear Reinforcement for All Heights of Fill

Provisions in LRFD related to shear reinforcement that are significantly different from those in Standard include a reduction in shear resistance provided by shear reinforcement of about 10 percent for LRFD, an increase in maximum spacing of shear reinforcement of about 40 percent for LRFD, and assuming f_c' is equal to 5 ksi, an increase in the minimum amount of shear reinforcement of 44 percent for LRFD.

In addition to the shear provisions listed above and regardless of the presence of shear reinforcement, LRFD requires a minimum amount of tensile reinforcement at all sections to satisfy the effects of combined tension and shear, and is in addition to requirements for minimum reinforcement and flexure. It is unlikely that this requirement will govern for typical culvert geometry and levels of reinforcement.

Crack Width Control

For buried precast culverts, LRFD and Standard define the crack width parameter "Z" as 100 k/in. and 98 k/in., respectively, which is a minor difference; however, LRFD does require a limit on total service load stress of 0.6 times f_y, which Standard does not. We believe the application of this provision to box sections was inadvertent and have neglected it in further comparisons of the two codes. In general, reinforcement is governed by crack width control for designs at depths greater than about 6 ft. The increase in "Z" from 98 k/in. to 100 k/in. will have a negligible effect on reinforcement requirements.

Requirements from both codes were developed primarily from testing of beams reinforced with deformed bars and as such make no distinction between crack width control properties of welded smooth wire fabric, welded deformed wire fabric, deformed bars or any reinforcement with stirrups. Heger, et al. [9] has shown that deformed reinforcement, or any reinforcement with stirrups have better crack control properties than welded smooth wire fabric, which in turn has better crack control properties than smooth wire with wide (greater than 8 in.) spacing of longitudinals. In contrast, both codes include the effect of reinforcement type for crack control for reinforced concrete pipe.

Fatigue

LRFD states that fatigue need not be investigated for buried concrete structures whereas Standard makes no distinction between buried culverts and other reinforced concrete structures and requires that fatigue be investigated.

Since box culverts with less than 2 ft of fill are designed with live load applied directly to the top slab, we do not consider these sections buried and recommend they be evaluated for fatigue. Both codes use the following equation to limit the stress range in reinforcement:

$$21 - 0.33\, f_{min} + 8\,\frac{r}{h} \tag{6}$$

where f_{min} =   minimum live load stress for fatigue load combination, include dynamic load allowance, positive if tension (ksi)

$\dfrac{r}{h}$ =   ratio of the lug base radius to the lug height radius for rolled-on transverse deformations and may be taken as 0.3 if this ratio is not known

A review of available information shows there has been little if any research performed on fatigue strength of welded smooth or deformed wire fabric. Since precast culverts commonly use fabric for reinforcement, and the shape of the rolled on deformations has a significant effect on fatigue strength [10], it is prudent, for both welded smooth and welded deformed wire fabric, to use r/h equal to 0.0.

LRFD provisions require that fatigue be investigated only for one design truck with an axle spacing of 32 ft, a load factor of 0.75, and a multiple presence factor of 1.0. The overall effect of these requirements, in spite of the generally larger dynamic load allowance factors and smaller strip widths for LRFD, show that fatigue governs reinforcement design more frequently for Standard than for LRFD. For typical culvert geometry with depths of fill ranging from 0 to 2 ft, fatigue require-ments may increase the reinforcement areas in the inside face of the top slab and bottom slab, and to a lessor extent in the outside face of sidewalls.

Deflection

For buried concrete culverts, LRFD provisions do not require investigation of deformations at the service limit state unless specifically required by the owner. In contrast, Standard provisions do not leave this decision up to the owner, but require that concrete structures have adequate stiffness so that deflections do not adversely affect the serviceability or strength of the structure. Both codes suggest a check on deflections is not warranted when member thicknesses meet the following requirement:

$$\text{Minimum member depth (in.)} \geq \frac{S + 10}{30} \geq 0.542 \text{ ft} \tag{7}$$

where   S   =   center to center distance between supports (ft), for culverts with haunches the face of support is located where the combined depth of the slab and haunch is 1.5 times the depth of the slab.

When deflection is investigated, both codes recommend the maximum allowable deflection be limited to span/800 for vehicular loading and span/1000 for vehicular and/or pedestrian loads, and the effects of reinforcement and cracking be considered when determining member stiffness. The intent of limiting deflection is to prevent potential deterioration of wearing surfaces and also to address the adverse psycho-logical reaction to flexible structures.

For longer span box culverts with relatively thin members, shallow earth/pave-ment cover, and truck loading, the above deflection limits may be exceeded. There is also a potential for differential displacement between adjacent culvert sections when only one side of a joint is loaded by a wheel. Significant displacement may unfavorably affect the design life of the wearing surface. For longer span culverts, we suggest that the code provision for minimum member thickness be considered unless calculated deflections are acceptable or some means such as shear connectors are provided to transfer load across joints.

Reinforcement Yield Strength

LRFD provisions allow for a yield strength of 60 ksi for Grade 60 deformed bars, 65 ksi for welded smooth wire fabric and 70 ksi for welded deformed wire fabric. Standard specifies similar strengths except that welded deformed wire fabric is limited to a yield strength of 65 ksi.

Comparison of Reinforcement Areas for Typical Designs

To illustrate the overall impact of the combined differences in code requirements, primary reinforcement areas for several combinations of culvert sizes and fill heights are evaluated for a design truck/HS 20 load based on provisions from LRFD, Standard and current AASHTO M 259 and M 273 Specifications. A typical reinforcement layout for box culverts is shown in Figure 3. The resulting reinforcement areas given in Figures 4, 5 and 6 show only sidewall outside face, top slab inside face and bottom slab inside face reinforcement areas and do not show areas typically governed by minimum steel requirements. Note that all culvert geometries are per M 259 and M 273 as appropriate. For the 4 ft span culvert in Figures 4, 5, and 6, this means the top slab is thicker (7.5 in.) for fill depths less than 2 ft than for depths 2 ft and greater (5 in.). Crack control is evaluated assuming a 4 in. spacing of reinforcement, the maximum allowed by M 259 and M 273. As noted previously, the LRFD limitation on total service load stress was not applied. Live loads are considered at all depths of fill, as assumed in the M 259 designs.

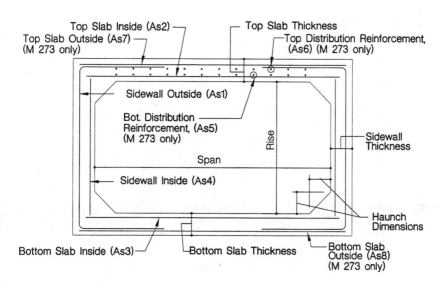

Figure 3 – *Reinforcement Layout for Box Culverts*

A comparison of the reinforcement areas in Figures 4, 5, and 6 show that for shorter spans with less than 2 ft depth, where top slab distribution reinforcement is present, reinforcement areas based on LRFD requirements are larger than those based on Standard requirements. The difference is most influenced by strip width provisions from LRFD. Other LRFD provisions that have a lesser effect on reinforcement areas include load factors, dynamic load allowance, live load surcharge and fatigue.

For depths of fill less than 2 ft, analysis shows that shear reinforcement is required for 4 ft and 8 ft span culverts designed in accordance with Standard provisions and 3 ft to 10 ft spans for LRFD provisions. The current AASHTO Material Standard for precast box sections (M 273) does not require stirrups for these conditions. However, Standard does allow designers to ignore the shear provisions for slabs when live loads are distributed using the strip width provisions. LRFD does not have a similar requirement.

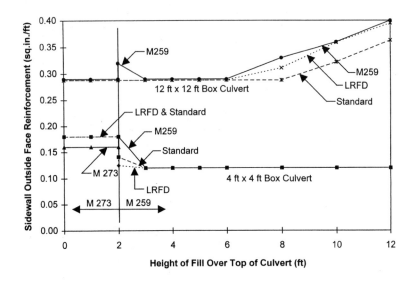

Figure 4 – *Sidewall Outside Face Reinforcement Areas*

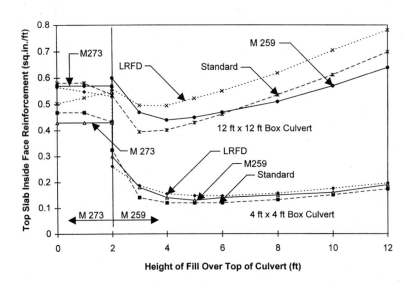

Figure 5 – *Top Slab Inside Face Reinforcement Areas*

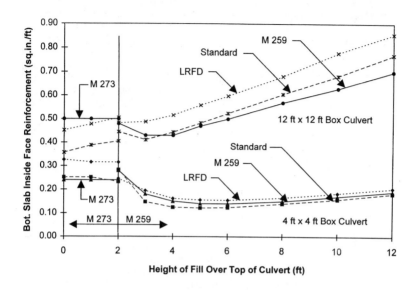

Figure 6 – *Bottom Slab Inside Face Reinforcement Areas*

At depths between 2 ft and 3 ft, where distribution reinforcement is not present in the top slab, differences in reinforcement areas are relatively small compared to those for the 0 to 2 ft fill range. For shorter spans, areas based on LRFD requirements are less than those based on Standard requirements. The difference in areas is mainly due to the fatigue requirements of Standard.

At fill depths above 3 ft, LRFD provisions typically result in larger reinforcement areas than those based on Standard provisions. The difference is mainly due to LRFD's load modifier for non-redundant earth load and live load distribution through earth fill.

Reinforcement areas from M 259 (depths of fill 2 ft and greater) and M 273 (depths of fill less than 2 ft), the predominant specifications used for design of box culverts, are presented in Figures 4, 5, and 6. A comparison of M 259 and M 273 with Standard and LRFD show similar trends; however, there are some notable differences in design requirements that sometimes affect reinforcement areas. The more important differences between M 259, M273, and Standard include:

- M 273 uses a minimum yield strength for reinforcement of 60 ksi as compared to 65 ksi for Standard. This effect decreases Standard reinforcement areas by about 8 percent.
- M 273 distributes the wheel load in the direction parallel to the span over a distance of 8 in. + 1.75 times the depth of fill. Standard does not allow any distribution of wheel loads in this direction and often results in larger reinforcement areas, especially for shorter spans. In addition, design to the Standard provisions occasionally result in high shear forces that exceed the diagonal tension strength if shear reinforcement is not provided.
- M 259 and M 273 do not account for the beneficial effects of compressive thrust, due to vertical live load and additional lateral earth load, on the flexural rein-

forcement in side walls and slabs, whereas Standard places no restriction on this beneficial effect. Additional compressive thrust typically reduces the flexural reinforcement in the sidewall outside face and inside face reinforcement, when not governed by requirements for minimum reinforcement. (See Fig. 4 for 12 ft x 12 ft boxes under deep fill.)

- M 259 does not check fatigue requirements as required by Standard. Fatigue requirements increase reinforcement areas in the 2 to 3 ft range of fill heights.
- M 259 and M 273 use capacity reduction factors for flexure of 0.90. Reinforcement areas in Figures 4, 5, and 6 for Standard use a capacity reduction factor of 0.95, although Standard allows the use of 1.0. This effect decreases M 259 and M 273 reinforcement areas by about 5 percent.

To demonstrate the overall effect, reinforcement areas based on LRFD and Standard provisions are compared to those from M 259 and M 273 in Table 4.

Table 4 – *Comparison of Reinforcement Areas*

Span x Rise (ft x ft)	Reinforcement Location	0 to 2 ft Cover		2 ft to 12 ft Cover	
		LRFD M 273	Standard M 273	LRFD M 259	Standard M 259
4 x 4	Sidewall Outside Face	1.13	1.13	0.95	0.96
	Top and Bottom Slab Inside Face	1.34	1.07	1.05	0.92
12 x 12	Sidewall Outside face	1.00	1.00	0.97	0.93
	Top and Bottom Slab Inside Face	0.99	0.91	1.16	1.02

Conclusions and Recommendations

A review of the LRFD and Standard codes show that there are several important differences in provisions relating to the design of buried precast reinforced concrete box culverts. Several combinations of box sizes and fill depths have been evaluated by the provisions of each code and show that LRFD provisions typically produce greater design loads and required reinforcement areas.

For fill depths less than 2 ft, the differences in reinforcement areas are most pronounced and are primarily caused by differences in provisions for strip widths. In addition, LRFD provisions require shear reinforcement for culvert spans up to 10 ft. Analysis based on Standard provisions also show that shear reinforcement is required for a similar range of spans but, provisions permit shear effects to be neglected.

Between fill heights of 2 ft and 3 ft, Standard requirements for fatigue, which are not present in LRFD, result in larger reinforcement areas. Beyond fill depths of about 3 ft, the difference in reinforcing areas decreases slightly with increasing depth with LRFD producing significantly larger areas. This difference is caused mainly by LRFD's requirements for distribution of live load through fill, load modifier for non-redundant earth load, and dynamic load allowance for depths less than 8 ft.

For depths of fill of about 6 ft and greater, crack width control governs the design of reinforcement. Differences in Standard and LRFD code provisions for crack control width have a negligible effect on reinforcement areas. It should be noted that we have not included the LRFD limitation of 0.6 f_y on total service load reinforcement stress, as we believe its inclusion was inadvertent.

A comparison of our findings based on LRFD provisions to those from M 259 and M 273 show that for depths of fill less than 2 ft, LRFD reinforcement areas are often significantly larger, and shear reinforcement is required by LRFD and not by M 273. Beyond 2 ft, areas based on LRFD are generally larger than those of M 259. LRFD's noticeably larger reinforcement areas and need for shear reinforcement for fill depths less than 2 ft seem to contradict the 20-plus years of satisfactory performance of culverts manufactured with reinforcement areas from M 259 and M 273 and suggests that LRFD provisions may be conservative. We feel that more study is needed to better define structural behavior as it relates to the distribution of wheel loads to slabs and through earth fills.

References

[1]   AASHTO, *LRFD Bridge Design Specifications*, 1st Edition, American Association of State Highway and Transportation Officials, Washington, D.C., 1994.

[2]   AASHTO, *Standard Specifications for Highway Bridges*, 16th Edition, American Association of State Highway and Transportation Officials, Washington, D.C., 1996.

[3]   Buckle, I. G., and Friedland, I. M., "Seismic Retrofitting Manuals for Highway Systems: System Assessment, Screening, Evaluation, and Retrofitting," *Proceedings of the Third U.S.-Japan Workshop on Seismic Retrofit of Bridges*, Dec. 1996, Public Works Research Institute, Tsukuba-shi, Japan, 1997, pp. 35-49.

[4]   James, R. W., "Behavior of ASTM C 850 Concrete Box Culverts Without Shear Connectors," *Transportation Research Record Report No. 1001*, Transportation Research Board, National Research Council, 1984.

[5]   Frederick, G. R., and Tarhini, K. M., "Model Analysis of Box Culverts Subjected to Highway Loading, Experimental Mechanics," *Proceedings of the 1988 SEM Conference on Experimental Mechanics*, Portland, OR, 1988, pp. 183-187.

[6]   AASHTO, "M 259-95, Precast Reinforced Concrete Box Sections for Culverts, Storm Drains, and Sewers," *Standard Specifications for Transportation Materials and Methods of Sampling and Testing*, 18th Edition, American Association of State Highway and Transportation Officials, Washington, D.C., 1997.

[7]   AASHTO, "M 273-95, Precast Reinforced Concrete Box Sections for Culverts, Storm Drains, and Sewers with Less Than 2 ft of Cover Subjected to Highway Loadings," *Standard Specifications for Transportation Materials and Methods of Sampling and Testing*, 18th Edition, American Association of State Highway and Transportation Officials, Washington, D.C., 1997.

[8]   Collins, M. P., and Mitchell, D., *Prestressed Concrete Structures*, Prentice Hall, Englewood, NJ, 1991.

[9]   Heger, F. J., and McGrath, T. J., "Design Method for Reinforced Concrete Pipe and Box Sections," report to American Concrete Pipe Association, Dec. 1982.

[10] Helgason, T., Hanson, J. M., Somes, N. F., Corley, W. G., and Hognestad, E., "Fatigue Strength of High-Yield Reinforcing Bars," *National Cooperative Highway Research Program Report 164*, Transportation Research Board, National Research Council, 1976.

James J. Hill, [1] John M. Kurdziel, [2] Charles R. Nelson,[3] and James A. Nystrom,[4]

Intelligent Technology for Concrete Pipe in the New Millennium

Reference: Hill, J. J., Kurdziel, J. M., Nelson, C. R., and Nystrom, J. A., **"Intelligent Technology for Concrete Pipe in the New Millennium,"** *Concrete Pipe for the New Millennium, ASTM STP 1368,* I. I. Kaspar and J. I. Enyart, Eds., American Society for Testing and Materials, West Conshohocken, PA, 2000.

Abstract: Precast concrete pipes were widely used even before the beginning of the twentieth century. By 1930, the currently used design and installation standards for concrete pipes had been empirically developed by Drs. Marston and Spangler. Nearly thirty years ago, the American Concrete Pipe Association began research on a more rational design method that evaluated the contributions of the strength of both the pipe and underlying soil. From that research, the design program Standard Installation Direct Design (SIDD) was developed.

In 1997, engineers at the Minnesota Department of Transportation successfully used the newly developed technology to update their standard practice for installing concrete pipe. A test, using native soils and simplified construction details, compared performance of the Marston-Spangler and SIDD installations.

Acceptance of direct design methods for concrete pipes is just the beginning of advancement of new technology for construction of pipelines. Rapid development of the internet, wireless communication links, geographical positioning systems, expert systems, and miniaturized testing equipment makes it possible for a manager of a construction project to integrate all phases of design, construction, and maintenance into a centralized information system. This paper describes how construction managers might use technology to mesh existing components of construction activities into an Intelligent Construction System.

Keywords: Concrete Pipe, Soil Structure, Direct Design, Intelligent Transportation Construction Systems (ITCS), Global Positioning Systems (GPS), Standard Installation Direct Design (SIDD), Soil Pipe Interaction Design Analysis (SPIDA)

[1] Civil Engineer, Minnesota Department of Transportation, 1500 West County Road B2, Roseville, MN 55113
[2] Civil Engineer, Hanson Concrete Products, 6055 150th St. West, Apple Valley, MN 55124
[3] Principal, CNA Consulting Eng., 2800 University Ave., Minneapolis, MN 55414
[4] Civil Engineer, The Cretex Companies, Inc., 311 Lowell Ave., Elk River, MN 55330

Introduction

At the turn of the last century, the first comprehensive procedures for the design, manufacture and installation of concrete pipe were developed. These procedures and methodologies were mainly the result of extensive research conducted by individuals such as Marston and Spangler.

In the early 1900s, a number of adverse field problems were being discovered with concrete and clay drain tile pipe. After a preliminary investigation, it became apparent to the researchers and designers of the time that there were not any means to determine loads on these structures, how to manufacture each pipe to a uniform strength designation, and what installation conditions should be used for bedding and backfilling. After decades of research testing and field verification, new design, manufacturing and installation standards were developed for these pipe. As a composite system, they became known as the Indirect Design Procedure or D-Load design.

This design methodology served the highway drainage, sanitary sewer and storm sewer users extremely well for over 100-years, but as with the last millennium, times necessitate the need for change. The driving factor for change in the next millennium is technology and optimization.

SIDD

In the 1970s, the American Concrete Pipe Association (ACPA) undertook a long-range research program to provide pipe designers with a better understanding of the behavior and interaction of buried concrete pipe and the surrounding soil. The result of the research concluded with a new method of concrete pipe design and installation which was named Standard Installation Direct Design (SIDD)[1]. The new installations were developed through actual field performance evaluations of soils and finite element soil-structure modeling with Soil-Pipe Interaction Design and Analysis (SPIDA). Further information on the development of SIDD can be found in ACPA's "Concrete Pipe Technology Handbook". The new installations are detailed for ease of construction as well as providing haunch support that reduces flexure in the invert of the pipe. The supporting stiffness of bedding material is based on objective criteria such as soil classification, placement and measured density.

SIDD provided the first major alternative for the design, manufacture and installation of concrete pipe since the adoption of the Indirect Design methodologies developed in the early 1900s. The SIDD designs provide not only a better engineered installation, but also a more cost effective alternative to the Indirect Design installations. In the spring of 1997, the Minnesota Department of Transportation (MnDOT), in conjunction with CNA Consulting Engineers and the Minnesota Concrete Pipe Association, initiated a research project to analyze the effectiveness and constructability of the SIDD pipe installation. The performance of the SIDD installations was compared to the Marston-Spangler installations, and as expected, the practicality and cost effectiveness of these installations were confirmed [2]. Further validity of these values has also been confirmed through a study comparing the U.S. Bureau of Reclamation's E's values to the performance of the SIDD installations [3].

The MnDOT research was not only very far reaching in its validation of the SIDD designs, but it also opened new doors for consideration in concrete pipe design, manufacture and installation. Technology was making it possible to obtain real-time information on the installation and performance of a pipeline. This research effort used a recently developed electromechanical instrument to directly measure the elastic modulus of the soil. The obtaining of the soil stiffness, as expressed by the elastic modulus, provided the key soil parameter for determining the effectiveness of the soil-pipe interaction for the SIDD methodologies [4]. A new era in evaluating pipeline design and installation was evolving.

The Transition

The SIDD beddings are only the first means of upgrading and modernizing reinforced concrete pipe design and installation. Technology has advanced to a point where it is now possible to customize pipe design, production and installation to not only minimize a project's overall cost but provide a complete new means for constructing a pipeline.

The SIDD design methodologies allow owners, producers and contractors to leverage their strengths. For owners, they can minimize a project's overall expense by balancing the costs of the pipe with different allowable beddings. The producer can supply product with the appropriate pipe reinforcing for the buried application as opposed to product manufactured to meet empirical testing criteria. The contractor can reduce installation costs by selecting bedding conditions or pipe strengths, which permit them to install pipe in less time and with fewer personnel.

Obtaining more detailed and instantaneous information on an installation allows for customization of a project. Soil types and conditions vary tremendously along the length of a project. If an owner and contractor know the quality of soil they are working with, installation procedures and pipe strengths can be modified accordingly when unstable soils are encountered or high quality in-situ soil conditions exist.

The MnDOT SIDD research project exposed a relatively old, standard construction practice for laying pipe to new innovative design and installation procedures. The use of electromechanical instruments for instantaneously measuring and monitoring the compaction efforts around a pipeline, definitively illustrated the actual level of variability of our soils and installation procedures for installing pipe. With this instant type of real-time data, poor areas of soil consolidation can be immediately corrected before any of the overburden is placed upon the pipe. We now have the technology to insure a high quality installation or at least know why we can not obtain it, given the site conditions.

This project also illustrated the opportunity to compare information obtained from the field with assumptions used in developing the initial design and planning the sequencing of the project. The research project was modified during the construction planning to obtain pertinent design field information on SIDD type of installations. The project was concentrated in the hands of a select number of individuals, who controlled the planning, design, manufacturing and construction. The limited size of this group enabled comprehensive decisions to be made in a very short period of time. The result was research data that was tailored to design assumptions.

In a macro-environment, the ability to obtain instantaneous information and modify project schedules, designs, manufacturing and installation is invaluable. This MnDOT research project demonstrated these benefits on a microenvironment. Technology is now available to apply these benefits to large-scale projects to increase construction productivity, construct an engineered installation and reduce the costs of project construction.

Project Development Cycle

Project owner's need to use and manage data more effectively for the planning, design, construction and maintenance of infrastructure systems. The development of a system to save the multi-functional information requirements of a project is key to operations in the next millennium.

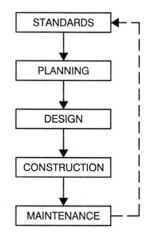

Figure 1. *Traditional Linear Construction Process*

The classic linear project cycle orients each operation of a project in a vertical stack (Figure 1) where each operation only interacts with the operation directly above or below them; standards-to planning-to design-to construction –to maintenance and then returning to standards.

This historical cycle worked well when time constraints and project costs were not an issue. As schedules have been tightened with cost conscious auditing, it has become necessary to conduct processes not in a linear series or even parallel, but concurrently. The level of competition has required the development of innovative systems and partnerships to compete for present and future work. Consolidated, interdependent companies have replaced the fragmented industry that existed in the past 100 years. The means for accomplishing these new project requirements and reacting to constantly changing performance requirements is embraced in the planetary system (Figure 2).

Under the planetary concept, all project development stages can be accomplished simultaneously. Design-build construction is just the latest example of how this concept operates. The project is actually still being designed as other sections of the project are under construction.

Maintenance has historically been the radial spoke municipalities ignore especially when cost constraints are a problem. Maintenance, however, is a critical component of this system. It determines how effective a product, process or construction method is performing relative to the other components on the project. Maintenance effects the other four stages just as each of them effect the others.

The key to the successful operation of the planetary concept is the center core, the main source of data of the information center. All project stages not only draw information from this core but also feed it. This interaction of information permits the efficient design of a project with regards to maintenance history and ease of construction. Planning of a project can be accomplished relative to pertinent and accurate standards substantiated through construction performance. Each spoke, therefore, relies on information from the other four spokes and feeds current and accurate data to each other through the main information depository at the core.

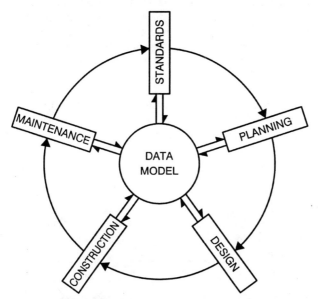

Figure 2. *Planetary Project Construction Process*

The center core is essentially a complex computer data base capable of being accessed at multiple levels: engineering, CADD standards, planning aerial photographs, construction soil borings, standard product dimensions and material properties, and maintenance estimated service life.

The engineering community has been using computer systems for design for decades, but the applications have always been specific and singular. The use of a complex, interactive, multiple level computer data system with composite core data to design, monitor and plan a project is relatively new. This computer system is not only the core to the planetary system for project development, but it represents a significant departure from the procedures and methodologies used for engineering design.

An Example of Construction of a Concrete Pipeline Using an Intelligent Technology System

Background

Intelligent Transportation Systems (ITS) are becoming well-established in cities around the world. Partnerships between citizens, government, and specialty equipment manufacturers are enhancing public transportation so it will be coordinated, integrated, operated and managed as a single system. As an example, in the Minneapolis-St. Paul,

Minnesota metropolitan area, a vast system of closed circuit TV cameras monitor traffic flow on major highways and feed information to a traffic control center. From there the rates of additional traffic entering the freeway system are regulated by entrance ramp meters. By controlling the rate in which the new traffic is permitted to enter the freeway system, traffic congestion at entry points is reduced and traffic already on the system is not impeded. Information from expert systems may be transmitted to message boards alerting drivers approaching entrance ramps of queue times and suggesting alternative routes. Some other systems are used to collect tolls or value pricing, provide travelers traffic information, help drivers adapt to speed limits, and provide priority at traffic signals for rapid transit buses.

At this time there has not been a similar development of an Intelligent Transportation Construction System (ITCS). An ITCS should be much simpler to develop because the owner of the system has the opportunity to define the organization and outcome of the system and, for most cases, the public is not directly involved. Most of the components necessary to create an ITCS exist today. Access to the Global Positioning System (GPS) has recently been given to the public. Cellular telephone and private communication satellite networks are becoming widespread with more capacity to transmit data. Remote PC users now can have secure wireless connections to computers in an information center. The Internet provides the common communication linkage.

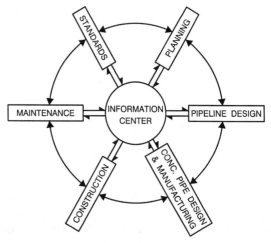

Figure 3. *ITCS Project Construction Cycle Demonstrating Construction of a Concrete Pipeline*

The design and construction of a concrete pipeline can provide an example of how a simple ITCS might be created. A schematic representation of a prototype construction system may be visualized by positioning the discrete design and construction operations in a circular pattern around an information center. Information can freely flow between operations through the information center. (Figure 3.)

All of the operations of the system will have access to the latest information and as construction progresses, compliance to standards may be monitored. Construction information will stream back to the original plan source creating an accurate as-built record.

Standards

Standards are the electronic archive of manufacturing and construction specifications and standard drawings that are necessary to complete construction of a project.

Planning

The planning operation of pipeline construction will provide documentation on description of the problem, the desired outcome, possible solutions and final implementation methods. Decision making would be aided by information provided by data bases indicating past and future land uses, topography and ecology, soil types, and service life. Planning information would be available in the information center to utilize in the pipeline designs.

Design of the Pipeline

The design of any pipeline includes a hydraulic analysis to calculate the volume and flow rates of a drainage system and a survey to determine the horizontal and vertical position of the pipeline. Electronic topographical databases can be integrated with hydraulic design programs to provide runoff information. Many computer aided drafting (CAD) programs include design features which may construct a model of the drainage area, calculate flow volume, select the concrete pipe size, and create an electronic document with all the design data. For this example, the SIDD design program will be incorporated to provide a structural pipe design for a specified type of installation and fill heights selected at critical locations along the pipe profile. All design criteria, special provisions, material quantities, precise global position coordinates, and concrete pipe strength requirements and manufacturing specifications will be a part of an electronic page of the construction documents. The concrete pipe supplier will be easily able to retrieve information for bidding and manufacturing pipes from the standards and design operations of the ITCS.

Design and Manufacture of Concrete Pipe

The concrete pipe supplier will have the option of manufacturing pipes with a standard design based on minimum installation requirements, or use an alternative design based on soil strength may be different from the original design. The standard design will include several variations in manufacturing based on reinforcing areas and concrete pipe wall thickness.

A new, more sophisticated finite element design model should be developed to replace SPIDA to provide better designs, particularly for trenches and provide the capability for virtual testing. A virtual test would electronically compare the design strength of a pipe to any load the pipe might be required to support.

Devices similar to surface imaging radar may be adapted to measure proper concrete cover and detect voids or hollow spots in the concrete. An electronic tracking device with encoded design and manufacturing information in it will be attached or embedded in the pipe.

Construction

The greatest economic benefit derived from the ITCS may occur in construction of a concrete pipeline. Heavy construction equipment control systems using GPS or laser beams will increase labor productivity and safety. In machine shops, computer aided manufacturing has been used for years to form complex parts on cutting tables or milling machines. The same electronic control principle may be applied to heavy construction equipment with satellite signals providing global coordinates for the horizontal and vertical alignment of the pipeline.

Researchers at North Carolina State University's Department of Construction Automation and Robotics Laboratory [5] have written numerous papers on robotic backhoe excavation and development of expert systems that would mimic an operator's response specific circumstances that occur while using a backhoe. Their research is directed at improvement in work place safety. By mechanically performing dangerous activities presently done by workers, both the safety and productivity of the crew will increase. In the summer of 1998, personnel of Robotics Laboratory conducted a full-scale test, on an actual construction project, that attempted to mechanically align, set, and join concrete pipe in a trench excavation.

The North Carolina State researchers used a manually aimed laser beam to provide geometric control for the line and grade of the pipeline. The backhoe was equipped with hydraulic pressure sensors and positioning devices on the arms that when mathematically integrated, calculated the breakaway force on the bucket. To integrate backhoe operation into an ITCS, additional electronic enhancements and linkage must be developed.

Electronic plans for the pipeline, including global coordinates for line and grade as well as the required soil strength, can be downloaded into an excavator equipped with an automated control system. The bucket of the excavator may be fitted with a precise GPS device that will transmit its location to the information center. The position of the bucket in the excavation can be limited to locations within an area indicated by global coordinates specified in the plans. The same positioning device may be used to ascertain that the location of the installed pipe meets plan requirements and transmit that position to the information center where it would become a part of the as-built records.

The information chip that is embedded in the concrete pipe will download manufacturing data through the GPS device to the information center. That information would assure the owner that the pipe meets standards, was installed in the proper location and inform the installer of the minimum modulus of elasticity of the soil used under the pipe invert and haunches. An electro-mechanical device with a GPS position transmitter mounted on a compactor will constantly monitor the soil modulus. When the proper soil stiffness is obtained, the device will indicate that the compactor may advance to another segment of pipeline. The GPS position transmitter will send the data to the information center and project monitors or inspectors.

Maintenance

Maintenance of a concrete pipeline is frequently performed by personnel who did not participate in the planning, designing or construction of the project, yet they are responsible for its successful operation. In an ITCS, the maintenance personnel would have direct access to the as-built information as well as the design and standards information. Data on the structural capabilities and location of each pipe on the job would be available. Standards, planning, designs, and construction methods could be evaluated and important information could be transferred back to appropriate operation of the construction process. Over a long period of time, the service life of the pipeline could be established and total maintenance costs accumulated to create a database for calculating life cycle cost schedules.

The ability to precisely locate buried structures would eliminate costly surveys required when surface transportation structures are reconstructed. Location data could be shared with public agencies responsible for locating underground utilities for other types of construction.

Control Center

The control center is the repository of all information generated by the construction of the concrete pipeline. The planners and designers may retrieve information from the standards. The pipe manufacturer can inquire about pipe quantities and requirements and have designs verified. The contractor will obtain project specifications and designs and during construction phase be able to transmit back geometric and structural data as the project is completed. Maintenance personnel will be able to use as-built data to keep the pipeline operating properly.

Conclusion

The description of an intelligent construction system used to design and install concrete pipe in a drainage project is easy, but is difficult to develop and implement. Drainage systems, usually installed in the initial phase of construction activities, are an economically small but essential component of a construction project. By focusing development efforts on drainage, a relatively simple aspect of construction, valuable insights can be learned for developing intelligence for even more complex construction projects. Nearly all the technology needed to implement an intelligent system to robotically install concrete pipe already exists in one form or may be readily invented. The biggest barrier to implementation is the lack of industry standards. There is no dominant leader in heavy construction, the engineering community, contractors, and construction material suppliers are all greatly fragmented. For intelligent construction systems to flourish, new standards for tests, testing apparatus, communication and software must be developed. At nearly every stage of describing the intelligent design and installation of concrete pipe, non-compatible systems with similar functions exist in the marketplace.

Examples:

Computer Added Drafting and Design (CADD) - Departments of Transportation use Intergraph™ software while most in private industry use Auto-Cad™.

Global Positioning Systems (GPS) - Both the United States and Russia have their own systems. By combining the two systems a civilian system with a high degree of accuracy may be provided. There are at least two systems for GPS correction base stations. A correction base station is necessary to provide global coordinates accurate enough for heavy construction.

Wireless Communication - Both analog and digital systems are available for voice communication but each having different coverage.

Computer Numerical Control for Excavating Machinery - Pipe will be installed to an alignment defined by global coordinates. To work within an intelligent system, all robotic controls for the machine will have to be able to read electronic plans.

Further development of concrete pipe soil interaction design methods will require new testing machines for the verification of design assumptions, and an analysis method that will utilize SPIDA or SIDD soil properties to model trench and embankment installations. Pipe designs will be subjected to a virtual test using an expert system that can simulate all possible conditions for a specific installation. Non-destructive test devices that measure concrete strength, cover over reinforcing, and steel reinforcing areas can be adopted from existing technology. Certification of the quality of the pipes may be encoded on an electronic device fastened to the pipe. New concrete pipe designs may be developed to withstand the rigors of robotic handling and lower soil structure installation standards.

References

[1] "Standard Practice for Direct Design of Buried Precast Concrete Pipe Using Standard Installations (SIDD)", ASCE 15-93

[2] "Evaluation of New Installations for Concrete Pipe", James J. Hill, John M. Kurdziel, Charles R. Nelson, and James A. Nystrom, American Society of Civil Engineers, *Pipelines in the Constructed Environment*, August 1998.

[3] "Replacing E' with the Constrained Modulus in Flexible Pipe Design", Timothy J. McGrath, American Society of Civil Engineers, *Pipelines in the Constructed Environment*, August 1998.

[4] "MnDOT Overload Field Tests of Standard and SIDD RCP Installations", J. Hill, J. Kurdziel, C. Nelson, J. Nystrom, M. Sontag, Transportation Research Board, January 1999.

[5] "CAD-Integrated Excavation and Pipe Laying", X. Huang and L.E. Bernold, Member, American Society of Civil Engineers, *Journal of Construction Engineering and Management,* September 1997

Innovative Case Histories

Timothy J. McGrath,[1] Ernest T. Selig,[2] and Mark C. Webb[3]

Field Tests of Concrete Pipe Performance During Backfilling

Reference: McGrath, T. J., Selig, E. T., and Webb, M. C., "Field Tests of Concrete Pipe Performance During Backfilling," *Concrete Pipe for the New Milliennium, ASTM STP 1368,* I. I. Kaspar and J. I. Enyart, Eds., American Society for Testing and Materials, West Conshohocken, PA, 2000.

Abstract: Full-scale field tests were conducted on the campus of the University of Massachusetts, Amherst, to evaluate the pipe-soil interactions that take place as pipes are buried and backfilled. The program included concrete, polyethylene, and corrugated steel pipe. This paper focuses on the data from the concrete pipe tests, and assesses it in light of the assumptions of the SIDD design approach adopted by ASCE and AASHTO.

Tests included two types of soil backfill, three compaction levels, three trench widths, varying haunching effort and one test with controlled low strength material. Eleven tests were conducted with 900 mm (36 in.) inside diameter pipe and three installations with 1500 mm (60 in.) inside diameter pipe. Primary measurements on the concrete pipe during backfilling and compaction included pipe-soil interface pressures, soil density, and soil stresses.

Results show significant variations in pipe behavior as a result of installation practices and generally confirm the assumptions of the SIDD design method. Compaction of backfill in the region from the springline to 45 to 60 degrees below the springline has a significant positive effect in mitigating poor bedding and haunching conditions, and the use of soft bedding is effective in reducing invert pressures on the pipe.

Keywords: reinforced concrete pipe, pipe, backfill, compaction, instrumentation

Introduction

The American Concrete Pipe Association's long-range research program of the 1970s and 1980s developed the current installation standards for reinforced concrete pipe

[1]Principal, Simpson Gumpertz & Heger Inc., Consulting Engineers, 297 Broadway, Arlington, MA 02474, tel. (781) 643-2000, e-mail: tjmcgrath@sgh.com.
[2] Principal, E.T. Selig, Inc., 105 Middle Street, Hadley, MA 01035, tel. (413) 586-1449, e-mail: selig@etselig.com
[3]Geotechnical Engineer, GFJ (Pty) Ltd., Consulting Engineers and Project Managers, P.O. Box 11449, Hatfield 0028 South Africa, e-mail: mcwebb@gfj.co.za

known as the SIDD method, for Standard Installation Direct Design [1-5]. These standards were developed to provide installation guidelines that reflect current practice for specifying backfill materials, and backfill compaction levels. The backfill requirements for SIDD embankment installations are presented in Figure 1 and Table 1.

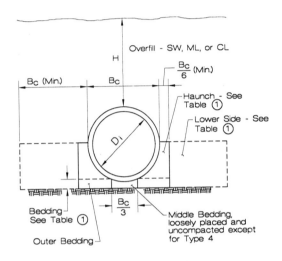

Figure 1 – *SIDD Embankment Installation Cross-Section*

Table 1 – *Backfill Requirements for SIDD Installations*

Installation Type	Bedding Thickness	Haunch and Outer Bedding	Lower Side
Type 1	For soil foundation, $B_c/600$ mm minimum, not less than 75 mm. For rock foundation, use $B_c/300$ mm minimum, not less than 150 mm.	95% SW	90% SW, 95% ML, or 100% CL
Type 2—Installations are available for horizontal elliptical, vertical elliptical, and arch pipe	For soil foundation, $B_c/600$ mm minimum, not less than 75 mm. For rock foundation, use $B_c/300$ mm minimum, not less than 150 mm.	90% SW or 95% ML	85% SW, 90% ML, or 95% CL
Type 3–Installations are available for horizontal elliptical, vertical elliptical, and arch pipe	For soil foundation, $B_c/600$ mm minimum, not less than 75 mm. For rock foundation, use $B_c/300$ mm minimum, not less than 150 mm.	85% SW, 90% ML, or 95% CL	85% SW, 90% ML, or 95% CL
Type 4	For soil foundation, no bedding required. For rock foundation, use $B_c/300$ mm minimum, not less than 150 mm.	No compaction required, except if CL, use 85% CL	No compaction required, except if CL, use 85% CL

As part of the pooled fund study *Fundamentals of Buried Pipe Installation,* sponsored by the Federal Highway Administration, National Science Foundation and eleven state departments of transportation, full-scale field tests were conducted on 900 mm (36 in.) and 1500 mm (60 in.) diameter reinforced concrete, corrugated steel, and corrugated polyethylene pipe [6-7]. The overall study was developed to investigate the behavior of buried pipe and backfill during the installation process. This paper reports the results of the tests on concrete pipe and findings concerning the validity of the SIDD assumptions for installation and backfill support.

Field Test Program

A total of 14 tests were conducted in different backfill materials and trench conditions, and with several bedding, haunching and compaction techniques. Eleven tests were conducted with 900 mm (36 in.) C wall pipe and three tests were conducted with 1500 mm (60 in.) B wall pipe.

Tests included three trench widths, with 300 mm, 600 mm, and 900 mm (12 in., 24 in. and 36 in.) at each side of the pipe. Full details of the tests are reported in [6-8]. Details of the tests on CLSM are presented in [9]. Details pertinent to this paper include:

- bedding was compacted for the full width of the trench for some tests and for others was left uncompacted directly under the invert,
- the two soil backfill materials were classified SW and SM (USCS classification system), one test was backfilled with controlled low strength material (CLSM), also commonly called flowable fill,
- three compaction conditions were used for each material: uncompacted, compaction with two passes of an impact tamper, or two passes with a vibratory plate compactor,
- haunching efforts consisted of rod tamping, shovel slicing and none.

The SW material was a 19 mm (3/4 in.), broadly graded crushed stone. The SM material was a poorly graded silty sand (silty sand). The CLSM backfill had a 28-day compressive strength of 779 kPa (113 psi).

In all cases the trenches were excavated with free standing vertical walls that were benched at approximately 1.2 m (4 ft) intervals. Figure 2 shows a typical trench configuration with the instrumentation. Instrumentation is described in detail in [6,7,10].

Tests were conducted at two sites. At the first site, called here the "sand" site, the soils were dense glacial deposits of coarse to medium sand (SP, SW-SM). The second site called the "clay site" consisted principally of a sedimentary varved clay deposit (CL).

The test procedure involved placing backfill in accordance with the test plan (Table 2) and reading instruments after every layer of backfill was placed. Lifts were about 300 mm (12 in.) thick after compaction. The final depth of cover over the test pipe was 1.2 m (4 ft) for all tests. At the end of a test, the site was immediately re-excavated to retrieve instruments and pipe and to inspect the condition of the bedding.

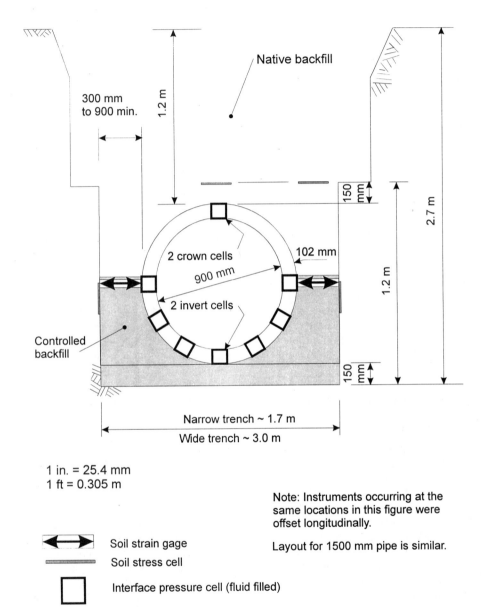

1 in. = 25.4 mm
1 ft = 0.305 m

Note: Instruments occurring at the same locations in this figure were offset longitudinally.

Layout for 1500 mm pipe is similar.

Soil strain gage
Soil stress cell
Interface pressure cell (fluid filled)

Figure 2 – *Cross-Section of Concrete Pipe in Trench with Instrumentation*

Table 2 – *Summary of Test Variables for Field Tests*

Test No.	Trench Width [1]	In situ soil	Pipe diameter mm (in.)	Backfill material	Sidefill compaction	Haunch [2]	Bedding compaction [3]
1	N	Sand	900 (36)	Stone	Rammer	SS	Fully compacted
2	N	Sand	900 (36)	Stone	None	N	Fully compacted
3	W	Sand	900 (36)	Stone	Rammer	SS	Sides compacted
4	W	Sand	900 (36)	Stone	Vibratory plate	N	Sides compacted
5	N	Sand	900 (36)	Silty sand	None	N	Fully compacted
6	N	Sand	900 (36)	Silty sand	Rammer	SS	Fully compacted
7	W	Sand	900 (36)	Silty sand	Vibratory plate	N	Sides compacted
8	W	Sand	900 (36)	Silty sand	Rammer	SS	Sides compacted
9	N	Clay	900 (36)	Stone	Rammer	SS	Fully compacted
10	N	Clay	900 (36)	CLSM	Rammer	--	Fully compacted
11	W	Clay	900 (36)	Stone	Vibratory plate	N	Sides compacted
12	N	Clay	1500 (60)	Stone	None	RT	Fully compacted
13	W	Clay	1500 (60)	Stone	Vibratory plate	RT	Sides compacted
14	I	Clay	1500 (60)	Silty sand	Vibratory plate	RT	Sides compacted

Notes: [1] N = narrow (O.D. plus 0.6 m, 24 in.), W = wide (O.D. plus 1.8 m, 72 in.), and I = intermediate (O.D. plus 0.9 m, 36 in.).

[2] SS = shovel slicing, RT = rod tamping, and N = none.

[3] Bedding was compacted with the vibratory plate. Fully compacted means the bedding was compacted over the full trench width. Sides compacted means that a strip directly under the pipe, one third of the pipe outside diameter in width, was left uncompacted.

If required by the test plan, all lifts were compacted with two coverages of the specified equipment (Table 2). The first lift over the top of the pipe was not compacted for a 300 mm (12 in.) width centered over the pipe. Resulting densities for each type of compaction were consistent (Table 3). In general water contents during compaction were dry of optimum. Only minimal effort was made to introduce moisture to improve compactibility, as this was deemed more closely related to actual practice. Moisture was added only when the material became dusty and difficult to work with.

Table 3 – *Soil Compaction Test Results and Moisture Contents*

Soil type	Compactor	Test Nos.	Compaction Test Results		Average Moisture Content, %
			Ave. % of Max. Unit Weight (AASHTO T99)	Stand. Dev., kN/m^3 (No. of test measurements)	
Stone	Rammer	1,3,9	92	0.5 (26)	2
	Vibr. plate	4,11,13	85	0.5 (14)	3
	None	2,12	79	0.4(8)	4
Silty sand	Rammer	6,8	95	0.2 (11)	8
	Vibr. plate	7,14	89	0.2 (13	7
	None	5	82	0.5 (6)	5

$1 \ kN/m^3 = 6.4 \ lb/ft^3$

Results

Measurements taken during the field test program covered a wide range of behavior.

Pipe-Soil Interface Pressures

The development of interface pressure on the concrete pipe for Tests 1 to 4, with stone backfill, and partial data for Tests 5 to 8, with silty sand backfill, are presented in Fig. 3. The end of test interface pressures for Tests 1 to 4 in a radial plot are presented in Fig. 4. In both figures the invert interface pressures are the changes after the pipe was set in place, thus the weight of the pipe is not reflected.

The highest invert pressure occurs for Test 2 where no haunching or sidefill compactive effort was provided but the bedding was fully compacted. Test 1, sidefill compacted with the rammer and haunched, and bedding fully compacted, shows a decrease in invert pressure as the sidefill was placed and compacted, suggesting that the

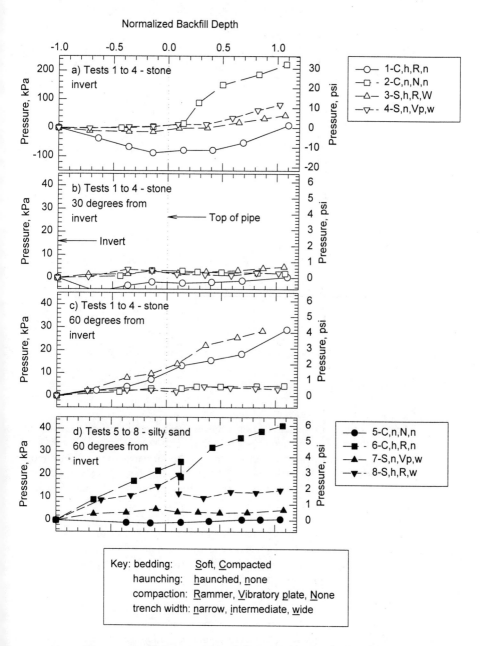

Figure 3 – *Concrete Pipe Interface Pressures during Backfilling*

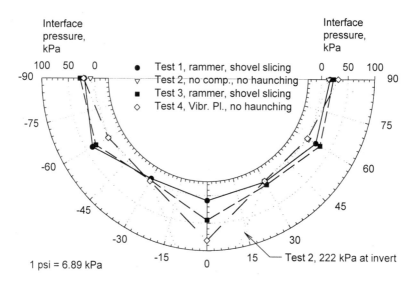

Figure 4 – *End of Test Radial Pressures on 900 mm (36 in.) Diameter Concrete Pipe, Stone Backfill*

compactive effort actually lifted the pipe off the bedding. Tests 3 and 4 show intermediate results.

Interface pressures at thirty degrees from the invert are low regardless of compactive effort or haunching effort. This suggests that design should always consider a region of the haunch as unsupported after backfilling.

The benefit of higher compactive effort is clearly seen in the interface pressures at sixty degrees from the invert. The two tests where the backfill was compacted with the rammer show high pressures. This is beneficial for pipe performance as it indicates more uniform support for the pipe. Interface pressures at this location for Test 4, compacted with the vibratory plate, showed very little difference from the pressures in Test 2, where no compactive effort was applied.

For Tests 5 to 8, with silty sand backfill (Fig. 3 (d), the data is similar to that for the tests with stone backfill (Fig. 3). The tests where the rammer compactor was used show higher interface pressures. Of interest are the drops that occur for Tests 6 and 8 at a backfill depth of about 0.1 m over the top of the pipe (Fig. 3(d)). This drop occurred overnight. The silty sand is sensitive to moisture, and the overnight delay in backfilling may have allowed the material to take up water and soften.

Interface pressure data for the other tests was similar. The end-of -test invert interface pressures under the 1500 mm (60 in.) diameter pipe (Tests 12 to 14, all with haunching), were between 100 and 200 kPa (15 to 30 psi), which were all less than the pressure under the concrete pipe in Test 2 without haunching.

Trench Wall Soil Stresses

Earth pressure cell data from Tests 5, 6, and 7 are presented in the form of stress versus depth of fill in Figure 5. Figure 6 shows the trench wall stress when the backfill was at the top of the pipe, and at the end of the test. In Test 5 with no compaction, only small lateral stresses develop at the springline level until the backfill level rises over the top of the pipe, and trench wall interface stresses are never greater than about 5 kPa (0.7 psi) during backfilling above the crown. For Tests 6 and 7 with compactive effort applied, horizontal stresses develop during compaction; and, as backfill is placed over the pipe, although the rate of increase in lateral stress at the trench wall is low.

The only direct comparison to evaluate trench wall stresses developed in narrow and wide trenches are Tests 1 and 3. In these tests the trench wall stress developed while placing the sidefill was greater for Test 3, the wide trench. The change in horizontal stress as the backfill was placed over the pipe was the same in Test 3 as in Test 1. The net effect was greater lateral stress when installed in the wide trench.

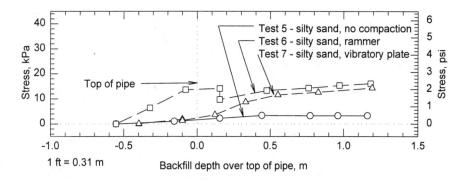

Figure 5 – *Horizontal Soil Stresses at Springline at Trench Wall-Backfill Interface*

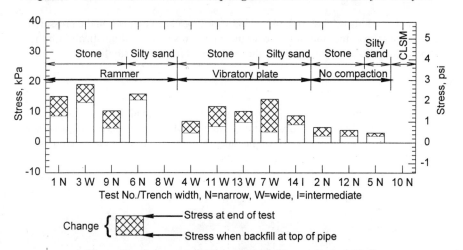

Figure 6 – *Summary of Horizontal Stresses at Trench Wall*

Vertical Soil Stresses Over Pipe

Vertical soil stresses directly over the pipe and over the sidefill at the completion of all backfilling, are summarized in Table 4 at the completion of all backfilling. The stresses are normalized by the geostatic soil stresses at the elevation of the gages based on the soil unit weights. The ratio of the crown to sidefill stress is not the arching factor, but is indicative of the arching of load onto, or off of, the pipe. No trend was noted based on diameter or trench width, thus the data is presented by type of compaction.

Table 4 – *Normalized Vertical Soil Stresses Over Test Pipe*

Compactor	Location	Concrete	
		Mean	Std. Dev.
Rammer (Tests 1, 3, 6, 8, 9)	Crown	0.96	0.10
	Sidefill	1.03	0.26
	Crown /sidefill (%)	94	
Vibratory plate (Tests 4, 7, 11, 13, 14)	Crown	1.04	0.08
	Sidefill	1.11	0.14
	Crown /sidefill (%)	94	
No compaction (Test 2, 5, 12)	Crown	1.28	0.23
	Sidefill	0.87	0.21
	Crown /sidefill (%)	147	

Table 4 indicates that for rammer and vibratory plate compaction the pressure is slightly greater over the sidefill than over the pipe. This reverses dramatically for the test with no compaction, because of the higher compressibility of the sidefill.

CLSM Backfill

The test with CLSM backfill was conducted with minimal instrumentation, thus the results are largely based on observation. The CLSM material was placed with ease and filled all voids in the lower half of the pipe. This test is discussed in much more detail in [9].

Analysis

Analysis of the field tests was undertaken with the finite element culvert analysis program, CANDE, Level 3 [11]. Complete finite element meshes were developed to represent the installation conditions of the tests [6,7].

Undisturbed in situ soils were modeled with estimated linear elastic properties while placed soils were modeled with non-linear behavior using the Duncan [12] hyperbolic Young's modulus with the Selig hydrostatic hyperbolic bulk modulus [13]. The CANDE User Manual, Appendix A, [11] contains two sets of Selig bulk modulus properties, called the "Modified," which are the defaults, and the "Hydrostatic," which must be input manually. All the major features of the tests such as, loose bedding and soft haunches, were included in the model as appropriate for any given test. Although the field tests were conducted to a depth of 1.2 m (4 ft) over the test pipe, the analyses were continued to a depth of 6.1 m (20 ft) to investigate implications of the various installation conditions under more demanding loading conditions.

The CANDE vertical and horizontal pressure distribution against the concrete pipe for Tests 1 and 2 are shown in Fig. 7.

Test 1, which was backfilled with stone, compacted with the rammer and haunched shows a variable vertical upward pressure distribution at the bottom even though haunched. This was borne out in the field tests by the low interface pressures measured at thirty degrees from the invert and the low penetration resistance measured after removal of the pipe. The vertical pressure distribution at the top of the pipe is relatively uniform at a depth of 1.2 m (4 ft), but shows a significant drop over the crown and over shoulder at a depth of 6.1 m (20 ft). This is likely the result of not compacting directly over the pipe. The low vertical pressures at the side of the pipe, are where the pipe wall is oriented vertically and are not significant The side pressure at the invert is low at all stages of backfilling; however significant pressures develop just above and below the springline. Note that the pressures in Fig. 7 are only changes in pressure due to fill over the crown, because the manner in which the CANDE analysis was conducted does not model compaction pressures.

Test 2, which was backfilled with stone, without compaction and without haunching showed a vertical pressure distribution at the bottom of the pipe that is peaked at the invert and does not develop the secondary pressure at the side of the pipe. This results from the lack of side support and haunching effort. At the top center, the vertical downward pressure distribution is uniform at all depths. The lateral pressure distribution at the side of the pipe is similar to that in Test 1, but lower in magnitude.

Measured interface pressures, and soil stresses at the trench wall and 150 mm (6 in.) over the crown are compared to the CANDE predictions in Fig. 8. The data presented is the change in interface pressure as the backfill was placed and compacted from an elevation 150 mm (6 in.) above the pipe, called the top of pipe, to 1.2 m (4 ft) above the pipe, called the end of test.

The CANDE predictions for invert interface pressure against the concrete pipe are consistently low relative to the field measured values, and the disparity increased as the compactive effort decreased (rammer, vibratory plate, none). The highest field change in invert pressure occurred in Tests 2 and 12 which had compacted stone bedding, no haunching, and no compaction. CANDE pressures were closer to the field values as the installation quality improved.

Interface pressures at the springline were quite low in both the CANDE analyses and the field data. The larger pressures developing above and below the springline, as shown in Fig. 7 indicate that the backfill is arching between the pipe and the trench wall, and little load travels directly through the backfill at the springline. This arching effect does produce substantial lateral support for the pipe.

Measured interface pressures at the crown of the concrete pipe were similar to those predicted by CANDE except for the tests with no compaction where CANDE underestimates the field values.

In general the CANDE finite element program provided quite good estimates of behavior and is quite powerful in its ability to address special design situations; however, the complexity of the program, and the uncertainty of actual installation conditions for most pipe, will probably result in CANDE being used only for special design situations. The use of the Selig "hyperbolic" bulk modulus values, the same as were used to develop the SIDD standard installations [2-4] produced a good match with the data.

Conclusions

The full-scale field tests conducted at the University of Massachusetts to study the influence of installation conditions on buried pipe behavior show that the methods used to achieve a specified installation condition can have a dramatic effect on the final behavior of the pipe. The tests validate the major assumptions of the SIDD design method adopted by AASHTO for the design and installation of concrete pipe. These assumptions include:

- soft bedding improves the overall pressure distribution by reducing the peak pressure at the invert,
- haunching is important in improving pipe support but never produces high pressures in the region about 30 degrees from the invert, and
- lateral pressures are significant, even in trench installations, when the pipe is installed with compacted backfill.

While equivalent pipe performance can be achieved with finer grained soils, the sensitivity to poor practice of installations with such backfill is increased, suggesting the need for greater quality control on the part of owners when such backfill materials are specified.

Pipe backfilled with CLSM performed well.

Acknowledgments

The tests reported in this paper were supported by funding from the National Science Foundation, the Federal Highway Administration, the states of California, Iowa, Kansas, Louisiana, Massachusetts, Minnesota, New York, Ohio, Oklahoma, Pennsylvania, and Wisconsin and by the Eastern Federal Lands Highway Division of the

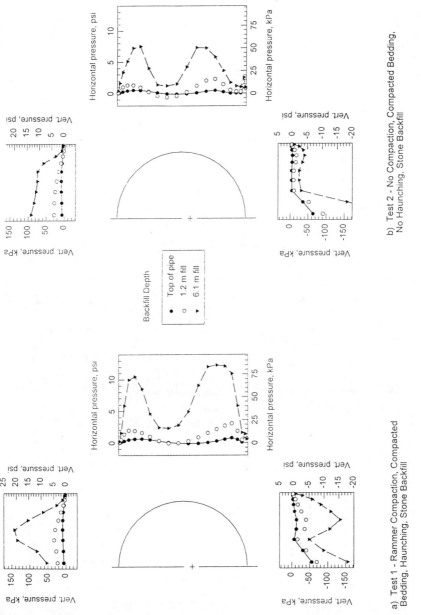

Figure 7 – *CANDE Predictions of Pipe-Soil Interface Pressures*

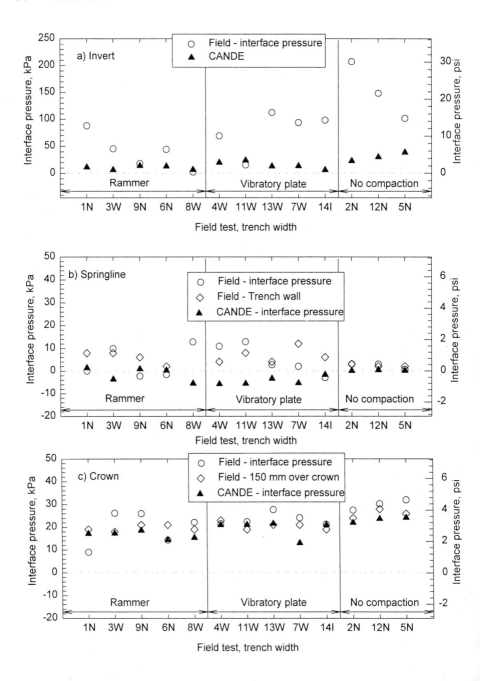

Figure 8 – *CANDE Interface Pressures Compared to Field Pressures for Concrete Pipe*

Federal Highway Administration. Concrete pipes used in the tests were donated by CSR/New England. Numerous useful suggestions were also offered by a number of consultants and representatives of various pipe suppliers as well, who, unfortunately cannot all be mentioned here. Massachusetts Highway Department generously provided a nuclear density gage for use during the field tests.

References

[1]   Heger, F.J., "New Installation Designs for Buried Concrete Pipe," *Pipeline Infrastructure - Proceedings of the Conference*, American Society of Civil Engineers, New York, NY, 1988.

[2]   Heger, F.J., Liepins, A.A., and Selig, E.T., "SPIDA: An Analysis and Design System for Buried Concrete Pipe," *Advances in Underground Pipeline Engineering – Proceedings of the International Conference*, American Society of Civil Engineers, 1985.

[3]   AASHTO Standard Specifications for Highway Bridges, 16th Edition, AASHTO, Washington, D.C., 1996.

[4]   AASHTO LRFD Bridge Design Specifications, 1st Edition, AASHTO, Washington, D.C., 1994.

[5]   AASHTO LRFD Bridge Construction Specifications, 1st Edition, AASHTO, Washington, D.C., 1998.

[6]   McGrath, T.J., *Pipe Soil Interactions During Backfilling*, University of Massachusetts, Ph.D. Dissertation, February, 1998.

[7]   McGrath, T.J., Selig, E.T., Webb, M.C., and Zoladz, G.V., *Pipe Interaction with the Backfill Envelope*, Report FHWA-RD-98-191, Federal Highway Administration, Washington, D.C., 1999. (This is the formal publication of Ref. 6, currently in progress.)

[8]   Webb, M.C., McGrath, T.J., and Selig, E.T., "Field Tests of Buried Pipe Installation Procedures," Transportation Research Board Annual Meeting, Jan. 1996; *Structures, Culverts, and Tunnels, Transportation Research Record*, No. 1541, National Academy Press, Washington, D.C., 1996, pp. 97-106.

[9]   McGrath, T.J., Webb, M.C., and Selig, E.T., "Field Test Installation of Buried Pipe with CLSM Backfill," *The Design and Application of Controlled Low-Strength Materials (Flowable Fill)*, ASTM STP 1331, A.K. Howard and J.L. Hitch, Eds., American Society for Testing and Materials, 1997.

[10] McGrath, T.J., Selig, E.T., and Webb, M.C., "Instrumentation for Monitoring Buried Pipe Behavior During Backfilling," *Field Instrumentation for Soil and Rock, ASTM STP 1358*, G.N. Durham and W.A. Marr, Eds., American Society for Testing and Materials, 1999 (in publication).

[11] Musser, S.C., Katona, M.G., Selig, E.T., *CANDE-89 User Manual*, Federal Highway Administration, Turner-Fairbank Highway Research Center, McLean, VA.

[12] Duncan, J.M., Byrne, P., Wong, K.S., and Mabry, P., "Strength, Stress-Strain and Bulk Modulus Parameters for Finite Element Analyses of Stresses and Movements in Soil Masses," *Department of Civil Engineering Report No. UCB/GT/80-01*, University of California, Berkeley, CA, 1980.

[13] Selig, E.T., "Soil Parameters for Design of Buried Pipelines," *Pipeline Infrastructure – Proceedings of the Conference*, American Society of Civil Engineers, New York, NY, 1988, pp. 99-116.

John J. Meyer[1] and Tim Whitehouse[1]

Case History of the Installation of a Sanitary Sewer Microtunnel Project

---

Reference: Meyer, J. J. and Whitehouse, T., **"Case History of the Installation of a Sanitary Sewer Microtunnel Project,"** *Concrete Pipe for the New Millennium, ASTM STP 1368*, I. I. Kaspar and J. I. Enyart, Eds., American Society for Testing and Materials, West Conshohocken, PA, 2000.

Abstract:

This Milwaukee Metropolitan Sewerage District project will require 3200 feet (975m) of 30 inch (750mm) diameter reinforced concrete microtunnel pipe to be installed at depths up to 64 feet (19m).
This specific design of microtunneling pipe was submitted to the owners by the low-bid installer and was accepted as an equal alternate to the original specifications. Microtunneling is a fairly new installation process, the technology of which allows for the trenchless installation of buried conduit while minimizing the need to disturb the street surfaces and/or private properties.
The project is scheduled to start June, 1998 and has an August, 1999 completion date.

Keywords: microtunneling, sewer construction

Project Description:

The Milwaukee Metropolitan Sewerage District authorized the construction of a 30 inch (750mm) diameter sewer in tunnel and open cut, and its associated manholes, to increase the Metropolitan Interceptor sewer capacity in southern Milwaukee. The project is designated the Ramsey Avenue Relief Sewer and was designed for the District by Thomas R. Wagner, P.E., of Kapur and Associates, Inc. The installation of the 3200 lineal feet (975m) of C76 Class IV reinforced concrete pipe by microtunneling actually commenced in early July, 1998 with projected completion being August, 1999.

---

[1]General Manager and Plant Manager, respectively, American Concrete Pipe Co., Inc. 5000 North 124th St., Milwaukee, WI 53225.

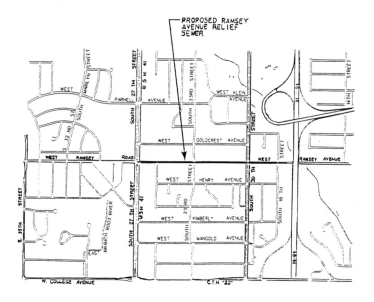

Figure 1 -- *Location Map*

A site location map, Figure 1, is included. The current sewerage drainage line flows to the south on South 27th Street into a drainage line on College Avenue and then east, eventually to the District's treatment (deeptunnel storage) plant. The installation of the Ramsey Avenue Relief sewer will bypass a majority of the 27th Street line to the College Avenue line and therefore allow for significant drainage contribution for areas now outside the current service area.

The installation method of tunneling was initially considered for this project because of topographic conditions. While depths of the line are 24 feet (7.3m) near Ramsey and S. 18th Street, and 23 feet (7.0m) at Ramsey and S. 27th Street, depths approach 64 feet (19.5m) for the "middle" run of the relief sewer near Ramsey and S. 23rd Street.

Site History

When the view of the project corridor, gained through visual inspection, was compared to a 1958 United States Geological Survey (USGS) topographic map, it appears that the project corridor has had little change except for an increase in residential properties scattered throughout the project corridor.

The land use in the vicinity of the project corridor is mainly residential in the east end of the project corridor with commercial properties surrounding the west end at the intersection of S. 27th Street and Ramsey Avenue. A series of figures (2-5) are included for the reader's review.

Figure 2 --*Residential land use at east end of project corridor.*

Figure 3 -- *East end of project corridor, looking west. Note first construction shaft.*

Figure 4 -- *Second construction shaft located adjacent to school.*

Figure 5 -- *Commercial land use at west end of project corridor.*

Site Geology and Hydrogeology

According to Soil Survey of Milwaukee and Waukesha Counties (USGS, 1971), the predominant soils in the project area are OUB2 (Ozaukee Silt Loam) and MtA (Houghton Muck) belonging to Ozaukee-Morley-Mequon Association and Houghton-Palms-Adrian Association, respectively.

In March 1997, Midwest Engineering Services, Inc. (MES) of Waukesha, Wisconsin completed seven soil borings at Crazy Jim's Auto located at 5839 South 27th Street as part of a combined Phase I and Phase II environmental site assessment for the site. Borings were advanced to depths of 11 to 21 feet (3.4 to 6.4m) below ground surface (bgs). Three borings were converted to groundwater monitoring wells. Based on the MES soil profile at this site, the soils in the project area generally consist of fill and possible fill soils consisting of brown and gray silty clays, silt, sand and gravel to depths of 6.5 to 10 feet (2.0m to 3.0m). Fill soils are underlain by black silty clays, greenish-gray clayey silt with sand seams and gray silty clays. Based on the groundwater measurements by MES from the existing monitoring wells located at the site, the depths to groundwater at the project corridor is expected to range from approximately 9 to 13 feet (2.7m to 4.0m) bgs. However, this depth is subject to change throughout the project corridor and is dependent on rainfall, surface run off, seasonality, and other environmental factors. Regionally, groundwater is expected to flow east/southeast toward Lake Michigan.

Site Assessment and Conditions

This site assessment has been performed in general accordance with the Environmental Reconnaissance and Record Search, Procedure 21-35-5 of the Wisconsin Department of Transportation (WisDOT) Facilities Development Manual. The on-site reconnaissance included visual observation for the presence of aboveground and underground storage tanks, septic systems, fill areas, depressions, distressed vegetation, and other indicators of potential environmental concern. In addition, interviews were conducted with some property owners. A site history evaluation and regulatory search was performed for the properties within 0.25 mile (402m) of the project corridor by interviewing local officials and reviewing state records, and EPA databases (NPL, CERCLIS, RCRIS, FINDS, and ERNS). The state records reviewed include several Wisconsin Department of Natural Resources (WDNR) and DILHR databases such as Underground Storage Tank (UST) Report, Leaking Underground Storage Tank (LUST) List, Hazardous Ranking List, and the Registry of Waste Disposal Sites In Wisconsin.

Based on the investigation procedure described, eight sites were identified as potential hazardous materials sites within 0.25 mile (402m) of the project corridor.

Site #1 is a WDNR LUST site. Remedial Site Investigation performed at the site indicated the presence of petroleum contaminated soils above the NR 720 standards that may extend past the south property boundary toward the West Ramsey Avenue right-of-way. Therefore, potential petroleum impacted soils may be encountered during excavation for sewer construction.

Site #2 is a former gas station and currently an automotive repair facility. This site is located adjacent to West Ramsey Avenue and is of concern because of possible migration of potential petroleum contamination from this property toward the Ramsey Avenue right-of-way.

The remaining six sites were reviewed and determined unlikely to impact the relief sewer construction.

Design, Material and Installation

The authors have chosen to relate to the audience the issues that were involved in the design of this relief sewer. The authors wanted to make the effort to emphasize the relationship of design, material, and installation involved in projects of this complexity. The authors have found that, on occasion, this interrelationship becomes forgotten and that a certain sector inadvertently receives forefront status. ASTM, as an organization, is comprised of producers, consumers, and general interest that have an appreciation for this type of triangular relationship.

Specifications and Testing

The material sector for this project is covered within the specification section of the Contract Documents. The design engineer used ASTM specifications as the base for design and testing to assure the District that they would be receiving the product that would meet the District's requirements. Pipe that is to be provided for the 410 lineal feet (125m) in open cut is thoroughly specified through the use of ASTM standards. The appropriate ASTM specifications address the wall thickness, the concrete strength, and the area, type, placement, number of layers and strength of the steel reinforcement, gaskets joint design, etc. Testing of randomly selected pipe is addressed through applicable ASTM specifications.

However, the reinforced concrete pipe that is to be provided for jacking needed to be modified beyond the requirements of ASTM C 76, Standard Specification for Reinforced Concrete Culvert, Storm Drain, and Sewer Pipe. The modifications required are necessary because C 76 is designed basically for shear (earth) load, not the axial load that is experienced by the pipe during the jacking operation. Fortunately, many designers, producers, and installers have experience with jacking pipe, and thus, as a team, they are able to provide significant knowledge of the necessary modifications. But once those modifications are incorporated into the finished pipe, the pipe is tested with methods similar to testing an open cut C 76 pipe.

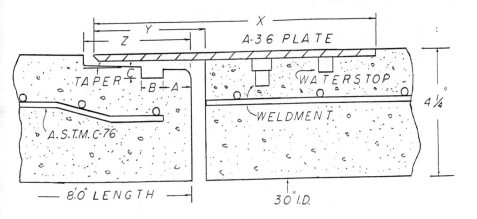

Figure 6 -- *30 inch (750mm) RCP Jacking Pipe*

By working with the low-bid installation contractor, the pipe manufacturer was able to design a pipe to meet the requirements of the contract documents and the needs of the contractor. The use of the structural requirements for C 76, as a minimum, allows the contract document requirements to be met. The next step was to determine the additional requirements of the contractor for the pipe's final design. Two critical items noted were having adequate pipe jacking capacity and the proper outside diameter to match the contractor's microtunnel machine. Concrete strength was increased from the required 4000 pounds per square inch (27.6MPa) of C 76 to 5400 pounds per square inch (37.3MPa) to give the pipe a jacking capacity of 390 tons (3470kN) including the projects required safety factor. This gave the pipe adequate strength to handle the jacking capacity of 300 tons (2780kN) delivered by the microtunneling machine used. The pipe was produced with an outside diameter of 38.5 inches (978mm) to meet the contractors overcut requirements for his machine which measured 39.4 inches (1000mm).

Installation

Figure 6 provides the reader with the basic conceptual overview of the jacking pipe that this being supplied to the jobsite. The pipe is best described as a 30 inch (750mm) C 76 Class IV pipe with a C wall and 5400 pounds per square inch (37.3MPa) compressive strength for the concrete. The pipe joint conforms to ASTM C 443, Standard Specification for joints for Circular Concrete Sewer and Culvert Pipe, using Rubber Gaskets. However, the reader should note that this is not a typical concrete joint. The joint consists of a joint ring that is machined from A-36 plate steel and is coated with two-part coat-tar epoxy. This special design of the joint provides the contractor a degree of "steerability" during installation in the type of ground conditions that are being encountered at the jobsite.

Figure 7 -- *Above ground set-up at second shaft adjacent to school.*

Figure 8 -- *Pipe positioned in jacking cradle at second shaft.*

Figure 9 -- *Close-up view of dewatering and slurry operations.*

Figure 10 -- *Located in the right half of the photo is one of the TV monitors which provides the operator in the control center continuous information.*

During the authors' most recent visit to the jobsite, the contractor was tunneling through what best could be described as wet sugar sand. This type of soil and the amount of dewatering were within the conditions described by the contract documents. The contractor's microtunneling machine and the tunneling pipe were designed (specifically for) these anticipated conditions.

As of January, 1999, the contractor has installed 2620 feet (799M) of pipe under this contract. The largest drive on this project was 676 feet (206M) followed closely by a push of 670 feet (204M). Both of these drives were completed with jacking pressures well under the design strength of the pipe and without the aid of intermediate jack stations.

The authors have included figures (7-10) for the reader's review. These figures concentrate on the tunneling operations set up in the shaft. The authors will continue to visit the jobsite until the projected completion in August of 1999.

Conclusion

This paper is being presented at a symposium entitled "*Concrete Pipe for the New Millennium*". With that theme in mind, the authors foresee the microtunneling installation process expanding in the new millennium and a growing need for ASTM standards for jacked/microtunneling pipe. Many unique designs for reinforced concrete pipe and the related components (i.e. steel, gaskets, joint design) are available in the marketplace but need a forum such as ASTM. ASTM provides an opportunity for standards to be reviewed and to be established. Heading into the new millennium, ASTM C-13 Committee is again positioned via its makeup of producers, consumers and general interest to address these subjects and bring forward to the marketplace products that will provide outstanding service well into the following millennium.

References

[1] Milwaukee Metropolitan Sewerage District, "Ramsey Avenue Relief Sewer", *Contract Documents,* Contract CO11GX010, November, 1997.

1395